CONTENTS

COLUMNS

22

Chris Schaie

PROJECTS

ON THE COVER:
Lulzbot's Taz 5 melted our minds with its high scores, great documentation, and open specs. Photo: Hep Svadja

24

James Burke

FAB FACTORY

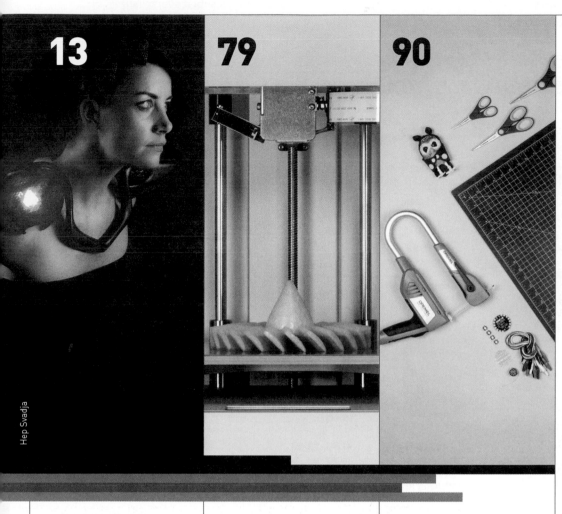

13

79

90

Hep Svadja

Issue No. 48, December 2015/January 2016. *Make:* (ISSN 1556-2336) is published bimonthly by Maker Media, Inc. in the months of January, March, May, July, September, and November. Maker Media is located at 1160 Battery Street, Suite 125, San Francisco, CA 94111, 877-306-6253. SUBSCRIPTIONS: Send all subscription requests to *Make:*, P.O. Box 17046, North Hollywood, CA 91615-9588 or subscribe online at makezine.com/offer or via phone at (866) 289-8847 (U.S. and Canada); all other countries call (818) 487-2037. Subscriptions are available for $34.95 for 1 year (6 issues) in the United States; in Canada: $39.95 USD; all other countries: $49.95 USD. Periodicals Postage Paid at Sebastopol, CA, and at additional mailing offices. POSTMASTER: Send address changes to *Make:*, P.O. Box 17046, North Hollywood, CA 91615-9588. Canada Post Publications Mail Agreement Number 41129568. CANADA POSTMASTER: Send address changes to: Maker Media, PO Box 456, Niagara Falls, ON L2E 6V2
. . . .
STATEMENT OF OWNERSHIP, MANAGEMENT AND CIRCULATION (required by Act of August 12, 1970: Section 3685, Title 39, United States Code). 1. Make: Magazine 2. (ISSN: 1556-2336) 3. Filing date: 10/1/2015. 4. Issue frequency: Bi Monthly. 5. Number of issues published annually: 6. 6. The annual subscription price is $34.95. 7. Complete mailing address of known office of publication: Maker Media, Inc., 1005 Gravenstein Highway North, Sebastopol, CA 95472. Contact Person: Kolin Rankin 8. Complete mailing address of headquarters or general business office of publisher: Maker Media, Inc., 1160 Battery St., Suite 125, San Francisco, CA 94111. 9. Full names and complete mailing addresses of publisher, editor, and managing editor. Publisher, Dale Dougherty, Maker Media, Inc., 1160 Battery St., Suite 125, San Francisco, CA 94111, Editor, Rafe Needleman, Maker Media, Inc., 1160 Battery St., Suite 125, San Francisco, CA 94111 , Managing Editor, Mike Senese, Maker Media, Inc., 1160 Battery St., Suite 125, San Francisco, CA 94111. 10. Owner: Maker Media, Inc., 1160 Battery St., Suite 125, San Francisco, CA 94111. 11. Known bondholders, mortgages, and other security holders owning or holding 1 percent or more of total amount of bonds, mortgages or other securities: None. 12. Tax status: Has Not Changed During Preceding 12 Months. 13. Publisher title: Make: Magazine. 14. Issue date for circulation data below: Oct/Nov 2015. 15. The extent and nature of circulation: A. Total number of copies printed (Net press run). Average number of copies each issue during preceding 12 months: 144,982. Actual number of copies of single issue published nearest to filing date: 145,657. B. Paid circulation. 1. Mailed outside-county paid subscriptions. Average number of copies each issue during the preceding 12 months: 64,156. Actual number of copies of single issue published nearest to filing date: 62,313. 2. Mailed in-county paid subscriptions. Average number of copies each issue during the preceding 12 months: 0. Actual number of copies of single issue published nearest to filing date: 0. 3. Sales through dealers and carriers, street vendors and counter sales. Average number of copies each issue during the preceding 12 months: 22,037. Actual number of copies of single issue published nearest to filing date: 16,250. 4. Paid distribution through other classes mailed through the USPS. Average number of copies each issue during the preceding 12 months: 0. Actual number of copies of single issue published nearest to filing date: 0. C. Total paid distribution. Average number of copies each issue during preceding 12 months: 86,193. Actual number of copies of single issue published nearest to filing date: 78,563. D. Free or nominal rate distribution (by mail and outside mail). 1. Free or nominal Outside-County. Average number of copies each issue during the preceding 12 months: 901. Number of copies of single issue published nearest to filing date: 825. 2. Free or nominal rate in-county copies. Average number of copies each issue during the preceding 12 months: 0. Number of copies of single issue published nearest to filing date: 0. 3. Free or nominal rate copies mailed at other Classes through the USPS. Average number of copies each issue during the preceding 12 months 0. Number of copies of single issue published nearest to filing date: 0. 4. Free or nominal rate distribution outside the mail. Average number of copies each issue during preceding 12 months: 9,477. Number of copies of single issue published nearest to filing date: 14,424. E. Total free or nominal rate distribution. Average number of copies each issue during preceding 12 months: 10,378. Actual number of copies of single issue published nearest to filing date: 15,249. F. Total distribution (sum of 15c and 15e). Average number of copies each issue during preceding 12 months: 96,571. Actual number of copies of single issue published nearest to filing date: 93,812. G. Copies not Distributed. Average number of copies each issue during preceding 12 months: 48,411. Actual number of copies of single issue published nearest to filing date: 51,845. H. Total (sum of 15f and 15g). Average number of copies each issue during preceding 12 months: 144,982. Actual number of copies of single issue published nearest to filing: 145,657. I. Percent paid. Average percent of copies paid for the preceding 12 months: 89.25%. Actual percent of copies paid for the preceding 12 months: 83.75%. 16. Electronic Copy Circulation: A. Paid Electronic Copies. Average number of copies each issue during preceding 12 months: 3,725. Actual number of copies of single issue published nearest to filing date: 3,668. B. Total Paid Print Copies (Line 15c) + Paid Electronic Copies (Line 16a). Average number of copies each issue during preceding 12 months: 89,917. Actual number of copies of single issue published nearest to filing date: 82,231. C. Total Print Distribution (Line 15f) + Paid Electronic Copies (Line 16a). Average number of copies each issue during preceding 12 months: 100,295. Actual number of copies of single issue published nearest to filing date: 97,480. D. Percent Paid (Both Print & Electronic Copies) (16b divided by 16c x 100). Average number of copies each issue during preceding 12 months: 89.65%. Actual number of copies of single issue published nearest to filing date: 84.36%. I certify that 50% of all distributed copies (electronic and print) are paid above nominal price: YES. 17. Publication of statement of ownership will be printed in the Dec/Jan 2016 issue of the publication. 18. Signature and title of editor, publisher, business manager, or owner: Todd Sotkiewicz, Business Manager. I certify that all information furnished on this form is true and complete. I understand that anyone who furnishes false or misleading information on this form or who omits material or information requested on the form may be subject to criminal sanction and civil actions.

76

Make:

EXECUTIVE CHAIRMAN
Dale Dougherty
dale@makermedia.com

CEO
Gregg Brockway
gregg@makermedia.com

CFO
Todd Sotkiewicz
todd@makermedia.com

EDITOR-IN-CHIEF
Rafe Needleman
rafe@makezine.com

EDITORIAL

EXECUTIVE EDITOR
Mike Senese
mike@makermedia.com

PRODUCTION MANAGER
Elise Byrne

COMMUNITY EDITOR
Caleb Kraft
caleb@makermedia.com

TECHNICAL EDITORS
David Scheltema
Jordan Bunker

EDITOR
Nathan Hurst

ASSISTANT EDITOR
Sophia Smith

COPY EDITOR
Laurie Barton

EDITORIAL ASSISTANT
Craig Couden

DESIGN, PHOTOGRAPHY & VIDEO

ART DIRECTOR
Juliann Brown

DESIGNER
Jim Burke

PHOTOGRAPHER
Hep Svadja

VIDEO PRODUCER
Tyler Winegarner

VIDEOGRAPHER
Nat Wilson-Heckathorn

MAKER MEDIA LAB

PROJECTS AND LAB DIRECTOR
Jason Babler
jbabler@makezine.com

PROJECTS EDITORS
Keith Hammond
khammond@makermedia.com

Donald Bell
donald@makermedia.com

LAB COORDINATOR
Emily Coker

LAB INTERNS
Paloma Fautley
Matthew Kelly
Anthony Lam
Adam Lukasik
Cameron Mira
Sandra Rodruigez

MAKEZINE.COM

DESIGN TEAM
Beate Fritsch
Eric Argel
Josh Wright

WEB DEVELOPMENT TEAM
Clair Whitmer
Matt Abernathy
David Beauchamp
Rich Haynie
Bill Olson
Susan Price
Ben Sanders
Alicia Williams

VICE PRESIDENT
Sherry Huss
sherry@makermedia.com

SALES & ADVERTISING

VICE PRESIDENT OF SALES
David Blaza
david@makermedia.com

SENIOR SALES MANAGER
Katie D. Kunde
katie@makermedia.com

SALES MANAGERS
Cecily Benzon
cbenzon@makermedia.com

Brigitte Kunde
brigitte@makermedia.com

STRATEGIC PARTNERSHIPS
Angela Ames

INSIDE SALES
Margaux Ryndak

CLIENT SERVICES MANAGERS
Mara Lincoln
Tara Marsden

CUSTOM PROGRAMS

DIRECTOR
Michelle Hlubinka

MARKETING

VICE PRESIDENT OF CORPORATE MARKETING
Vickie Welch
vwelch@makermedia.com

MARKETING PROGRAMS MANAGER
Suzanne Huston
suzanne@makermedia.com

DIGITAL MARKETING COMMUNICATIONS MANAGER
Brita Muller
brita@makermedia.com

MARKETING EVENTS MANAGER
Courtney Lentz
courtney@makermedia.com

MARKETING SALES DEVELOPMENT MANAGER
Jahan Djalai

BOOKS

PUBLISHER
Brian Jepson

EDITORS
Patrick Di Justo
Anna Kaziunas France

MAKER FAIRE

PRODUCER
Louise Glasgow

PROGRAM DIRECTOR
Sabrina Merlo

MARKETING & PR
Bridgette Vanderlaan

SPONSOR RELATIONS MANAGER
Miranda Mota
miranda@makermedia.com

COMMERCE

GENERAL MANAGER OF COMMERCE
Sonia Wong

SENIOR BUYER
Audrey Donaldson

RETAIL CHANNEL DIRECTOR
Kirk Matsuo

E-COMMERCE MANAGER
Michele Van Ruiten

INVENTORY PLANNER
Percy Young

ASSOCIATE PRODUCER
Arianna Black

CUSTOMER SERVICE

CUSTOMER SERVICE REPRESENTATIVES
Ryan Austin
Camille Martinez

Manage your account online, including change of address:
makezine.com/account
866-289-8847 toll-free in U.S. and Canada
818-487-2037,
5 a.m.–5 p.m., PST
cs@readerservices makezine.com

PUBLISHED BY

MAKER MEDIA, INC.
Dale Dougherty

Copyright © 2015 Maker Media, Inc. All rights reserved. Reproduction without permission is prohibited. Printed in the USA by Schumann Printers, Inc.

CONTRIBUTING EDITORS
Stuart Deutsch, William Gurstelle, Nick Normal, Charles Platt, Matt Stultz

CONTRIBUTING WRITERS
John Abella, Josh Ajima, Samuel N. Bernier, Tom Burtonwood, Chandi Campbell, DC Denison, Matt Griffin, Shawn Grimes, Kurt Hamel, Kacie Hultgren, Jason Loik, Claudia Ng, Jim Rodda, Luis Rodriguez, Chris Schaie, Madelene Stanley, Vishal Talwar, Chris Yohe, Spencer Zawasky

Comments may be sent to:
editor@makezine.com

Visit us online:
makezine.com

Follow us on Twitter:
@make @makerfaire @craft @makershed

On Google+:
google.com/+make

On Facebook:
makemagazine

CONTRIBUTING ARTISTS
Gunther Kirsch, Rob Nance

ONLINE CONTRIBUTORS
Cabe Atwell, Philip Bradley, Gareth Branwyn, David Calkins, Jon Christian, Emily Coleman, Jeremy Cook, Jimmy DiResta, Miriam Engle, Adam Flaherty, Leia Gatch, Faine Greenwood, Karen Hickman, Grady Hillhouse, Jess Hobbs, Heather Kleiner, Rachel Kohout, Meredith Lee, George LeVines, Sandeep Mistry, Goli Mohammadi, Haley Pierson-Cox, Tony Sherwood, Taylor Soule, Theron Sturgess, Andrew Terranova, Madison Worthy

CONTRIBUTORS

What's one thing you bought at the store, that you'd like to make (or print) instead?

Matt Stultz
Warwick, Rhode Island [3D print editor]
I've got a green 1982 Puch Maxi Luxe moped. I would love to print my own parts, to keep the vintage beauty running.

Chandi Campbell
Springfield, Missouri [A RepRap Family Tree]
I recently bought a little Chevy truck, and I'm very happy with it, but I really look forward to building my own open source electric vehicle.

Chris Schaie
San Diego, California [CNC Mechanical Iris]
Definitely my tool chest. I will build my own some day. I will … I will, I swear it.

Gunther Kirsch
Sebastopol, California [3D print photographer]
Ice cream, definitely ice cream. Having it 3D printed would be even better! I'm not sure how well it would print, but I'd eat it anyway.

Kacie Hultgren
New York, New York [Laser cutter reviews]
I wish I could 3D print big! I'd go crazy designing full-size furniture for my apartment.

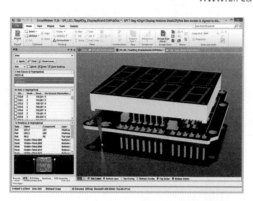

DIY INTEL EDISON MIDI SYNTHESIZER

Written by DJ Hard Rich

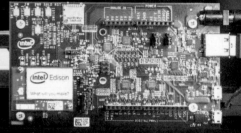

Unleash your inner DJ with this custom synth using the **Intel Edison** module

I'VE BEEN EXPERIMENTING WITH THE INTEL EDISON TO SEE HOW I CAN USE THIS SUPER-SMALL DEVICE AS THE COMPUTATIONAL BRAINS OF MY CUSTOM AUDIO PROJECTS. My latest build is a MIDI synthesizer, which replaces bulky equipment and performs just as well, if not better, than some commercial options I've seen. Plus, I made it myself. Here's how I did it.

1. FLASH THE EDISON MODULE

» Follow the Intel guides to set up the Edison module at http://intel.ly/1Drs15a.
» Ensure the Yocto image you use to flash your module is ww25-15 or later.
» While you're at it, insert the USB hub into the Edison with Arduino breakout board's USB socket.
» Then attach your USB sound card and MIDI keyboard to the USB hub and your physical build is done.

2. LOG INTO THE SYSTEM

» Using a terminal emulator such as Putty for Windows or Screen for OSX and Linux, connect to Edison module by following the OS specific instructions.
» For Windows enter in the serial port associated with the Edison module — such as COM1 — and set the baud rate at 115200; Linux uses /dev/ttyUSB0 with the same baud rate.
» For OSX open a terminal and type, sudo screen -L /dev/cu.usbserial hit tab to complete, space and 115200. My full command reads as sudo screen -L /dev/cu.usbserial-AJ035QGP 115200.
» At the login prompt, type root and hit return.

3. CONFIGURE THE SYSTEM

» At the command prompt, type configure_edison --setup.
» Follow the onscreen instructions to create a root password, unique hostname, and Wi-Fi setup.

4. SETUP DEBIAN CHROOT ENVIRONMENT

» From the command line on the Edison

```
    Starting Int
  ] Started Intel_X
    Starting Network
    Starting Getty on
[  OK  ] Started Getty on tty
    Starting Serial Gett
[  OK  ] Started Serial Getty
[  OK  ] Reached target Login
    Starting HSU runtime
[  OK  ] Started Network Name
   OK  ] Started File System
   OK  ] Started HSU runtim
    Mounting /factor
  ] Mounted /facto
    6243] syst
```

2

module type, wget https://cdn.makezine.com/make/Edison_debianscripts.zip. This downloads two shell scripts you will run to setup a Debian environment on your Edison module.

» Unzip the archive by typing unzip Edison_debianscripts.zip.

» Type in mount /home/ -o dev,remount and press return. This command handles mounting the home directory for the chroot Debian environment.

» Run the two shell scripts by typing, sh bootstrap-debian-edison.sh. This will take a bit of time to run, since it's pulling a barebones Debian Linux Operating System.

» Then execute the second shell script to handle mounting partitions. Type sh setup-chroot-mounts.sh.

5. CHROOT TO DEBIAN AND INSTALL PACKAGES

» "Change root" into the Debian root directory using the command, chroot /home/root/debian /bin/bash.

» Create a home directory for the root user in your Debian chroot by typing mkdir /home/root.

» Next type in apt-get update; apt-get install fluxbox tightvncserver qjackctl amsynth dbus-x11 alsa-utils to install the necessary software for the project.

» When prompted if you want to enable realtime process priority, respond yes.

6. EDISON VNCSERVER SETUP

» From the Debian chroot environment, start the VNCserver daemon by typing, dbus-launch vncserver :1.

» Follow the onscreen prompts and enter a VNC server password — it doesn't have to be different than your Linux login password, but it's not a bad idea to make them different.

» When prompted if you want to make the password view only, respond no.

7. VNC TO THE MODULE

» On your Edison, type ifconfig wlan | grep inet and write down the inet addr, such as 10.2.2.154. This is the IP of the Edison module's Wi-Fi chip. You'll need this address to know where to point your VNC viewer.

» If you don't have one already, install a VNC

viewer on your laptop such as RealVNC. http://bit.ly/1EONwmM, though any VNC viewer will do.

» Run your VNC viewer and enter in the Edison module's IP address you wrote down plus :1. Here's how mine looks, 10.2.2.154:1.

» Give it a moment and you should see your Edison module's desktop with the classic Debian swirl logo.

8. AUDIO APPLICATIONS

» Logged in via VNC, right-click anywhere on the desktop and select JACK Control to launch the JACK Audio Connection Kit. Go to setup and make sure your USB Audio Device is selected, then click OK to start the JACK server.

» Right-click again on the Desktop and select amsynth from Applications/Sound to start the synthesizer.

9. SSH AND ACONNECT

» At the command prompt, type aconnect -o which spits back all the valid MIDI outputs connected to your Edison device — specific outputs will depend on the sound card and MIDI device you use.

» Note the two numbers associated with your MIDI keyboard and type in the following command: aconnect xx:0 129:0 where xx is the device number output by aconnect -o.

This MIDI project is just the start. Add a touchscreen and custom case to make this project truly portable. Or use the Edison module's mini breakout board, add a battery and embed the build inside your keyboard. Make it your own! Then get ready to rock. ✪

Maker Inspiration, Lessons, and Tips

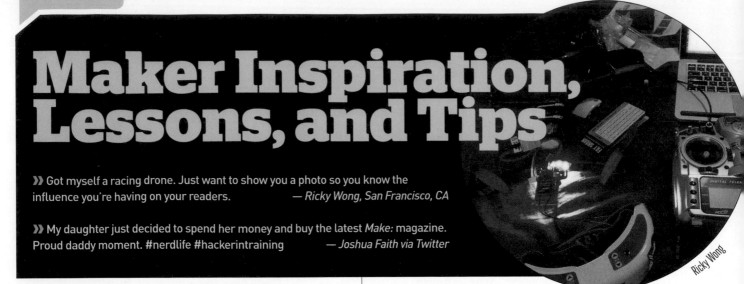

» Got myself a racing drone. Just want to show you a photo so you know the influence you're having on your readers. — *Ricky Wong, San Francisco, CA*

» My daughter just decided to spend her money and buy the latest *Make:* magazine. Proud daddy moment. #nerdlife #hackerintraining — *Joshua Faith via Twitter*

Ricky Wong

IN RESPONSE TO "BUILD A PVC WATER BALLOON CANNON" FROM MAKE: VOLUME 46, PAGE 80,

makezine.com/go/water-balloon-cannon

» I modified the plans by using 1.5" PVC for everything, but then used a reducer/expansion joint to connect to a 3" diameter pipe so we could load larger balloons. We also drilled in an air compressor fitting instead of a valve stem and pumped it up to about 100psi and can shoot a larger balloon a couple hundred feet easy. We built it for an engineering demo for a group of middle schoolers and they loved it! Thanks for the plans. — *Aaron Watwood, via the web*

» This canon exceeded all my expectations. If you get the balloon just the right size, add the perfect amount of water, and perhaps go above 30psi, you are talking at least 350 feet. This was my first project out of *Make:* magazine. Thanks for the article and the summer fun! — *Jonathan Bechtel, via the web*

» When I was a "kid" we did this same thing with a spud gun. Instead of water to lubricate the barrel we would squeeze the balloons and wrap them with a piece of newspaper to make them cylindrical/sausage shaped, effectively squeezing a larger balloon into a smaller barrel and minimizing friction between balloon and barrel. We could consistently launch water balloons 100–150yds and hit unsuspecting pedestrians. They often never knew what or who hit them. — *Lee Brock, Okinawa, Japan*

(Make: *warns: If you over-pressurize your cannon, the PVC can explode and shatter — and PVC shrapnel doesn't show up on X-rays.*)

IN RESPONSE TO "6 LESSONS FROM ROBOT COMBAT,"

makezine.com/go/six-lessons-from-robot-combat

» You missed the most important lessons from robot combat. Here's what I took away from my years in the sport, going back to pre-*BattleBots* in the late '90s:

- Have fun. There will come a time when you're knocked out, broken, or just plain broke. If you can't sit back and enjoy it like a spectator with an all-access pass, you're doing it wrong.
- Never stop learning. Grab new skills. Your new friends will help.

- Make friends with everyone involved. You're only competing inside the ring — outside you're all the same flavor of nerd with the same weird hobby.
- Share your experiences. Become the teacher you wish you'd had when you were a newb.

Robot combat changed the whole trajectory of my life. Early interest in it informed some of my studies in college. Many years of post-college bot building taught me all the stuff I missed when I was IN college. I ended up stumbling into an actual robotics job, becoming a robotics mentor to a dozen kids, and generally became the "Maker" I am today. Better yet, I'm still friends with many of the early bot builders nearly two decades later. — *Mike Herbst, San Diego, CA*

READER TIPS

IN RESPONSE TO "SIX 'NOW, WHY DIDN'T I THINK OF THAT?' SHOP TIPS," makezine.com/go/six-shop-tips

» I like the tips. But as a 40+ mechanic I suggest automotive valve grinding compound to increase the coefficient of friction on any slippery metal, bolt, nut, and especially screws. Outside a rounded nut or bolt, a little dab will make the wrench grip like it is welded. The compound uses diamond dust and is $2 at most auto parts stores. It even makes vise grips grip tighter. — *Jim Bob via the web*

» Add baby powder to super glue to instantly cure it. Put the powder inside a cleaned out glue bottle or something else to control the flow since it can be messy if you're not careful. — *James Rossfeld, Torrance, CA*

MAKE AMENDS

In Volume 47's "New in the Shed" (page 90), the Othermill should have been credited to Other Machine Co., not Otherlab. Other Machine Co. manufactures the Othermill and has been independent from Otherlab since 2013. Thanks to Other Machine Co.'s Ezra Spier for setting us straight.

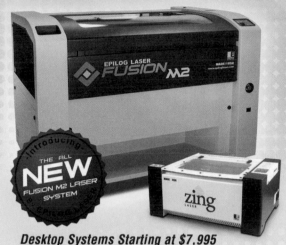
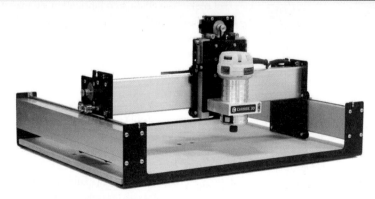

Gunther Kirsch

Beautiful Obsessives
Welcome to the rabbit hole

BY RAFE NEEDLEMAN, editor-in-chief of Maker Media.

HI, I'M RAFE, THE NEW GUY. I just started as editor-in-chief of *Make:* (and makezine.com). I'm not sure I can express how happy I am to be here, and how much *Make:* means to me. But I'm going to try.

Of the many jobs I've had in technology journalism, the most interesting (until now) was writing a daily column about startup companies in the first dot-com bubble, from 1998 to 2001. I covered an entrepreneur a day. Each of them dreamed of changing the world. A few of them actually did.

I loved covering startups and their founders because the best founders are maniacally obsessive, and obsessive people are interesting. If you sit down and really talk to somebody who is completely immersed in a topic, and you keep asking them to go deeper into the details, obsessives get happier as the discussion gets more technical. It is a beautiful thing to see a person — any person — really groove on a topic, and it's fascinating to see the depth of knowledge a person can have.

It's why I love Makers and making. The level of knowledge that a master Maker can have on a topic is just immense. The level of craftsmanship that a person can bring to even a seemingly trivial project is fascinating. For example, you could just weld metal. Or you could make welding into an art (see makezine.com/go/nice-welds). You could splice two wires together. Or you could get into a heated

argument on the best way to make a splice (makezine.com/go/nasa-splice). Win or lose the argument, it doesn't matter. You learn a lot, and you improve your craft.

So given any chance to dive down a rabbit hole of making, crafting, and fabricating, I'll take it. Because I think it's cool to see people work on things they love.

Speaking of skillfully using tools, the issue in your hands is our annual look at 3D printing. We assembled more than a dozen experts on 3D printing and fabricating in our lab over the course of a long weekend, and let them go deep into a selection of the best 3D fabrication tools.

This year, for the first time, we included not just standard filament fusing 3D printers, but also some mini CNC mills and a few laser cutters. And then there's my favorite 3D fabricator, Printrbot's massive Crawlbot, a router rig that can literally pull itself across a 4×8 sheet of plywood to "print" furniture pieces. It's spectacular.

Putting this issue together was a treat. Most of us, in our daily lives, rarely see so many people come together to share their expertise as openly as our testers did for this issue. It was great fun. We all learned a lot about the craft of 3D fabrication, and a lot about the state of the art for 3D printing, too.

We fussed and argued over every printer we got into the lab, and every test we ran, and every result. And now you have the outcome of our obsession in your hands. I hope you use it to feed your own. ⏣

Nick Strayer, Evernote

A Cut Above

Glowforge's Dan Shapiro believes a laser cutter should live beside your 3D printer

Maker Pro
Q&A

Written by DC Denison

DAN SHAPIRO IS DETERMINED TO REINVENT THE LASER CUTTER. He is CEO and co-founder of Glowforge, which he calls "the first 3D laser printer." With $9 million already raised, the company garnered an additional $5 million in sales in just the first week of a 30-day, half-price preorder promotion for the $4,000 machine this September. Housing a 40W laser (45W "pro" edition available), Glowforge packs a wallop, cutting plywood and acrylic up to ¼" thick, and features a cloud-based software package that may finally make laser cutting intuitive.

Previously, Shapiro created Robot Turtles, the best-selling board game in Kickstarter history, which teaches programming fundamentals to preschoolers. He has also served as CEO of two earlier startups, experiences that have informed his recent book, *Hot Seat: The Startup CEO Guidebook*.

This is your third time as a founder and CEO. What's changed because of the Maker revolution?

For a long time, the path to go from inspiration to product was incredibly short for software and atrociously difficult for hardware. In the past five years there's been a revolution. Suddenly hardware's become achievable.

A huge part of that is the amazing work that the 3D printing community has done, a huge part is the work that pioneers on platforms like Arduino and Raspberry Pi have done.

Once it starts to become easier, more people do it, and as more people do it, it becomes easier. So we're in this magnificent feedback loop where the Maker community is building the bridge as they're walking across it.

Has the definition of "homemade" changed?

Yes, we're reinventing homemade. We're not just making things, we're making things that make things. And we're doing that because we want to upend the notion of what it means to make things.

The Industrial Revolution brought us low cost and speed and convenience — which are not to be trifled at — but at the cost of losing quality and customization, and the longevity of the product. I don't want to take mass production and put it in my house.

What I want is to combine all the lost benefits of "homemade" with all the value that we got with low cost and convenience.

Plastic was the star of the 3D printing revolution. Will low-cost laser cutters showcase other materials?

Yes. When I think about the products that I use every day and that I care about, somewhere between few and none of them are made entirely of plastic. Most of the products that make up my day come from sheet goods. I'm standing in my office now looking at my desk, which is made of plywood, looking at my bag which is made out of leather, looking at my jacket, fabric — all goods that are made out of sheets, or assembled from precision-cut pieces.

As a three time startup veteran, do you have any advice for Makers who are thinking of going pro?

It's about starting at the end: figuring out who is going to be as excited about what you are doing as you are.

The reason Robot Turtles was such a success was that it was a thing the world wanted to exist that did not yet exist. When I asked, "Who wants to teach their preschooler to program?" 13,000 hands shot up in the air and said, "Yeah!" That was what pulled a successful product out of what I was trying to do. I didn't have to push something that I was excited about to an apathetic audience.

Figure out who will pull the wonderful product out of you, rather than how to push the product you care about to an audience. ◐

For more Maker Pro news and interviews, visit makezine.com/go/maker-pro, and subscribe to the Maker Pro Newsletter at makezine.com/go/maker-pro-newsletter.

DC DENISON is the editor of the *Maker Pro Newsletter*, which covers the intersection of Makers and business. He is the former technology editor of *The Boston Globe*.

6 Awesome Things to 3D Print

Start your printers – these projects are amazing and the 3D files are free!

Written by *Make:* Editors

A

SNAP-TOGETHER R/C CAR

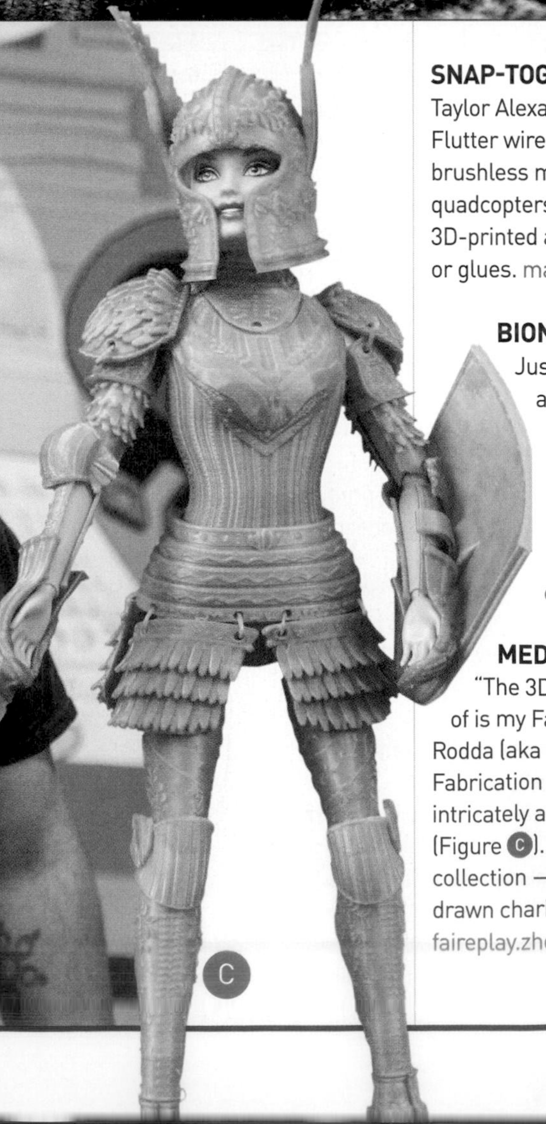

Taylor Alexander built this little ripper around his Flutter wireless robotics boards and a powerful brushless motor (Figure Ⓐ) like you'd find in quadcopters. His ingenious design is totally 3D-printed and snaps together without screws or glues. makezine.com/go/flutter-scout

BIONIC CLAWS

Just flex your forearm and a fraction of a second later your 4-inch Wolverine claws fly out! Brian Kaminsky 3D printed this awesome project (Figure Ⓑ) to demonstrate the possibilities of his company's MyoWare muscle sensor. makezine. com/go/3d-printed-bionic-claws

MEDIEVAL BARBIE ARMOR

"The 3D modeling work I'm most proud of is my Faire Play Barbie Armor" says Jim Rodda (aka Zheng3), one of our *Make:* Desktop Fabrication Shootout testers (see page 25). It's intricately authentic-looking, and super badass (Figure Ⓒ). And that's just the start of the collection — you will not believe the crazy cat-drawn chariot your battlin' Barbie can ride in. faireplay.zheng3.com

B

C

A & B: Hep Svadja, Jim Rodda, Steve Jurvetson, Joao Duarte, Daren Banarsë

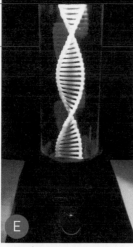

TechFab Fusion
The 21st century apparel maker

Written by Mike Senese and Nathan Hurst

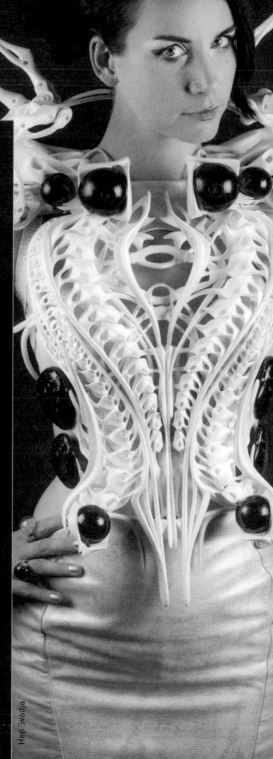

ANOUK WIPPRECHT HAS A UNIQUE APPROACH TO 3D-PRINTED FASHION, melding technology with a healthy dose of physicality. It's about art and creativity as much as wearables or electronics. Her Spider Dress (made in conjunction with Intel) and Smoke Dress (made for Audi) have established her as an important contributor — and in-demand personality — among forward-thinking fashionistas.

WHAT'S YOUR DESIGN PROCESS?
I start on paper and then trace and morph in Photoshop before importing to a 3D modeling program like Maya or Rhino. I use ZBrush for surfaces, and finish with a rendering program like KeyShot to get a visual of the 3D data.

HOW DO YOU DETERMINE FIT?
I scan my clients' bodies to not only have their sizes but also their posture and bone structure in 3D data. Knowing how a person stands, moves, and is built is really important for a perfect design. I use 123D Catch, an Artec hand scanner, or Intel's RealSense built into my design tablet.

DO YOU CREATE MANY ITERATIONS?
A minimum of three. The first, you print your pieces way too bulky because you are afraid they'll break; the second, way too small and detailed so they indeed do break; on the third you hit a good balance.

ANY 3D DESIGN ADVICE?
Don't be afraid of the unknown. Try as much as you can. ●

Hep Svadja

SUPERSONIC ROCKETS

We've featured Steve "The Rocket Man" Jurvetson before: VC funder, rocket maker, and collector of Apollo trinkets. His new obsession is 3D printing model rockets (Figure **D**). He's painting PLA fins with 5-minute epoxy and they're surviving Mach 2. Awesomeness. flickr.com/photos/jurvetson/16243380206

DNA LAMP

Want a new twist on the lava lamp? Joao Duarte at eLab Hackerspace in Portugal designed this stunner from scratch in TinkerCAD and printed the helix in glow-in-the-dark filament, then added Arduino-sequenced LEDs and a rotating base for an endless-stair-climb illusion that's mesmerizing (Figure **E**). instructables.com/id/3D-Printed-DNA-Lamp

NEOTRADITIONAL MELODICA

Forget toy melodicas and make this real reed instrument. Daren Banarsë's "world's first" 3D-printed melodica (Figure **F**) sounds great and looks lovely, with optional ivories cut from salvaged piano keys. melodicaworld.com/3d-printed-melodica-world-first

Make more great 3D-printed projects at makezine.com/projects.

11 Rad Things to CNC

Fire up a CNC router or laser cutter to make these impressive, freely shared projects

Written by *Make:* Editors

FULL CUSTOM ELECTRIC GUITAR

Steve Carmichael used his X-Carve CNC router to shape this entire hardwood guitar body, neck, fretboard, and inlays (Figure Ⓐ). He calls it "a great example of combining the accuracy of the X-Carve with the finesse of handwork."
inventables.com/projects/electric-guitar

CHAIRS AND WORKBENCHES

The plywood Fabchair (instructables.com/id/fabchair) by Russian designer Yaroslava is a simple slot-together seat that's at home in a makerspace, playroom, or classroom (Figure Ⓑ). And check out SketchChair's free software (sketchchair.cc) — you just draw a chair-shaped squiggle and it's automatically turned into a unique, custom-fit design you can CNC.

While you're at it, give your machines a place to sit: design your own CNC Maker Bench (Figure Ⓒ) using AtFab's parametric program, and build it at makezine.com/go/cnc-maker-bench.

PRO-GRADE YO-YO

Industrial design student (and ex-*Make:* engineering intern) Eric Chu milled this ball-bearing yo-yo (Figure **D**) in Delrin resin on a desktop Othermill. "It's a great test-part," he says. "The smoother the yo-yo spins, the more accurate the CNC is!" instructables.com/id/OtherYo

GORGEOUS 3D MAPS

Josh Ajima, another of our *Make:* Digital Fabrication Shootout testers, converted a topographic map of the Chesapeake Bay watershed into this museum-quality display (Figure **E**) by laser-cutting nine layers of colored cardstock and stacking them for brilliant relief. You can cut it in acrylics or woods instead (and 3D print it too). thingiverse.com/thing:908712

DINOSAUR SAFETY GEAR HANGERS

Make: engineering intern Sam DeRose designed these simple, eye-catching racks for your ear and eye protection (Figure **F**), laser-cut from acrylic in four flavors: *Tyrannosaurus rex*, *Velociraptor*, *Brontosaurus*, and *Triceratops*. makezine.com/go/workshop-laser-cut-dinosaur-safety-gear-holders

SEAHORSE CORBELS

Intricate architectural details are easy to carve on a CNC. We love these coral-reef corbels by Mike Tyler (Figure **G**), cut on a ShopBot Buddy, for

decorative shelf supports, porch posts, or any corner that needs a little undersea wonder. shopbottools.com/files/Projects/Seahorse_Corbel_Tutorial.pdf

MAKERSPACE SHEDS

Shelter 2.0 is a flat-pack, easy-to-assemble 10'×16' shed for small makerspaces; grab the files for cutting at shelter20.com (watch for a new mini version coming soon), then build it at makezine.com/go/build-a-makerspace.

For more headroom, make our soaring 14'×16' CNC Makerspace Shed (Figure **H**) by Rick Schertle and Lendy Dunaway, big enough to build an art car or teach a soldering class. makezine.com/go/cnc-makerspace-shed

QUICKLAP CANOE

When boatbuilder Bill Young bought his first ShopBot router he immediately set about redesigning the traditional lapstrake canoe, each special plank shape cut precisely by CNC and then epoxied and fiberglassed (Figure **I**). He's been doing special projects for ShopBot ever since. 100kprojects.com/project_pages/Quicklap_Canoe

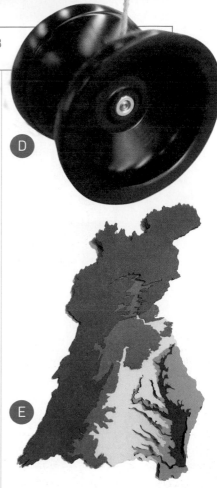

Make more great CNC projects at makezine.com/projects.

Supersized
Seven-Segment Clock

Time Required: `A Few Hours` **Cost:** `$60-$80`   Written by Matt Stultz

3D-print this desktop LED clock with jumbo digits that glow from within

MATT STULTZ
is *Make:*'s 3D printing editor and leader of this year's Digital Fabrication Shootout.

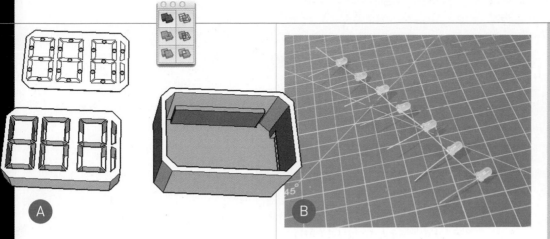

IF THERE'S ONE THING NONE OF US HAVE ENOUGH OF, IT'S TIME. Tracking the time we do have has been a goal of many great inventors — and today, making your own clock is practically a rite of passage for Makers.

In this project, I'll show you how to use a 3D printer to build your own jumbo-sized seven-segment LED desktop clock. It glows beautifully from within, and it's even more enjoyable when you tell your friends you made it yourself.

THE CLOCK THAT CONQUERED THE WORLD

By the 1300s, we were creating accurate mechanical clocks with the invention of the escapement. But even the best pendulum clock would lose time aboard a rocking ship, rendering it useless for navigation. In 1714 the British government offered a reward to anyone who could devise an accurate shipboard clock, and in 1761 John Harrison's marine chronometer finally met the test — over the course of 10 weeks it was only off by 5 seconds.

The first electric clock would be patented in 1840 and the first quartz-timed clock built in 1927. Advances in solid-state electronics in the 1980s made it possible for quartz LED clocks to spread throughout the world. Mechanical clocks live on as decorative or luxury time pieces (a Rolex still has gears), but the humble seven-segment LED clock is now everywhere: your car, your cable box, your microwave. Let's make one.

1. PREPARE THE 3D PARTS

I started this project by designing the three printed parts in SketchUp (Figure A): the case, the LED holder, and the digits block (the display face). Printing the parts takes about a day on most desktop 3D printers.

I drew the digits to resemble a standard seven-segment display. The thin front of the display diffuses the light from the LED within, causing the

entire segment to glow. You could redesign the segments for a different look, but be sure to leave plenty of space between each segment to help reduce the bleed between them. The segments are sized to be large enough for the LEDs to fit into them but still hold them in place with a snug friction fit.

Test-fit your LEDs into their holes in the holder. If they won't go in, use sandpaper or a knife to carefully enlarge the holes. You want them to be in there tight, so don't go overboard.

2. SOLDER THE LED CHAINS

Hold an LED with its leads pointing straight up, the longer lead toward you and the shorter lead away from you. Bend the short (negative) lead of your LED over 90° to the left, near the base of the LED. Do this to all 23 of your LEDs.

Now get your soldering iron fired up. Take 2 of your prepared LEDs, and solder the tip of one of the bent leads to the corner bend of the other. Take a third LED and solder its corner bend to the unsoldered negative tip of the previous pair. Continue doing this until you have a chain of seven LEDs (Figure B).

Make 2 more chains like this.

3. INSTALL THE LEDS

Place the 3D-printed LED holder on top of the digits block, with the "8s" to the left and the "1" to the right. Now insert an LED chain into the leftmost "8," starting with the top leftmost hole, so that the chain forms a backward S shape. Do the same to your other two "8s" with your other two LED chains (Figure C).

Place your last two LEDs into the last 2 holes for the "1" digit. Turn the bent leads toward each other and twist them together leaving a ⅛" (3mm) tail sticking up, and then solder the wires together (Figure D).

Clip all of the long (positive) leads from your LEDs down to about ⅛" (3mm).

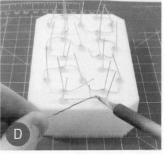

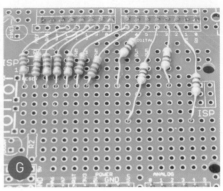

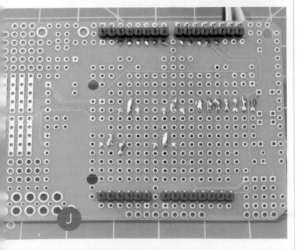

4. POPULATE THE PROTO SHIELD

Now you'll prep your prototyping shield. Start by soldering on the header pins. I find the easiest way to do this is to first insert them into my Arduino (Figure **E**), then set the proto shield on them (Figure **F**). This way even if the headers on your Arduino aren't perfectly straight, your shield will be matched to the Arduino.

Remove the proto shield from the Arduino. On the shield, place the seven 220Ω resistors into the holes corresponding to digital I/O pins 6 through 12, then place their opposite ends into a straight row of holes in the "blank space" on the shield. Do the same for the 4.7K resistors in holes 2 through 5, but this time leave a bit of space for the transistors (at least 1 column and 3–4 rows) as shown in Figure **G**. Ensure that none of the resistor leads are touching each other.

With the flat side of the transistors facing toward you, insert them in the board one row below the 4.7K resistors, with their center pins aligned with the resistor lead. Solder them in place (Figure **H**).

5. CONNECT THE CABLES

Cut 8" of ribbon cable and remove any extra wires so you're left with only 8 wires. On one end of the cable, cut 1½" off 7 of the leads, leaving just one long wire. Separate all these wires by about ½" and strip them all about ⅛". On the other end of the cable, separate all the wires by about 1" and strip them all.

Starting with the short wire furthest away from the long tail, solder it into the hole below the resistor for pin 12. Work across all of the short wires for pins 12–6. Solder the long wire into the proto board to correspond to the left pin of the transistor that's hooked

up to pin 2 of your Arduino (Figure **I**). As you solder these wires in place, create solder bridges that connect them to their appropriate leads above (Figure **J**).

Prep 2 more cables identical to the first, and solder them directly below, again connecting to pins 12–6, and to the left pins of the transistors for pin 3 and 4, in that order.

Prep a final cable in the same fashion, but instead of having 8 wires, this one should only have 3. Solder it into pins 8 and 7, and then into the left pin of the final transistor hooked to pin 5.

Finally, cut 4 lengths of hookup wire long enough to stretch from the right pins of the transistors to a place on the board where you can bring them all to ground (Figures **K** and **L**).

6. CONNECT THE LEDS

Before soldering, slip a length of heat-shrink tubing over each wire in your cables.

Starting with the topmost cable on the shield and the leftmost "8" digit in your pattern of LEDs, solder the cable to the now short and yet unsoldered LED leads using the guide here (Figure **M**). Solder the remaining wire (from the pin 2 transistor) to the remaining tail of the LED chain (Figure **N**).

Connect the 2 remaining 8-wire cables, moving from top to bottom, to the 2 remaining "8" digits, moving from left to right. For the final, 3-wire cable, connect its long (transistor) wire to the tail created by the 2 clipped leads tied together on the "1" digit, and connect its 2 short wires to the remaining 2 short leads from the 2 LEDs.

With that, your soldering work is done! I'd recommend you follow Steps 7 and 8 to test your clock before finally shrinking the tubing in place.

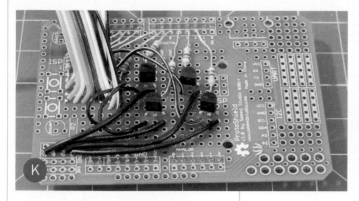

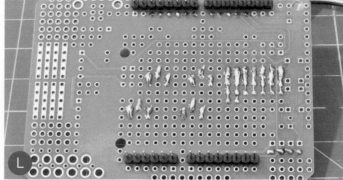

7. UPLOAD THE ARDUINO CODE

Install the 2 included libraries into your Arduino environment. (If you've never done this before, follow the guide at arduino.cc/en/guide/libraries#toc5.) These libraries make it easy for you to keep track of time and to work with seven-segment displays like the one you just finished building.

Now plug your proto shield and a USB cable into your Arduino, and your Arduino into your computer. If you've never used this particular Arduino before, you might need to install some drivers before you can proceed. (If this is a problem, consult arduino.cc for help.)

With your Arduino plugged into your computer, open *ThreeDPClock.ino* in the Arduino IDE. On the Tools→Board menu, select the Arduino you're using, and on Tools→Port select the appropriate serial port. Then hit the Upload button to load the code onto your Arduino.

8. SET THE TIME

Sadly, computers are not great at dealing with dates we understand; they work better on their own date/time routines. To set the time and date on your clock, you need to set the *epoch time* (aka *Unix timestamp*) on your Arduino. Go to epochconverter.com and use the drop-down menus to set your desired time, then hit the "Human date to Timestamp" button. Copy the "Epoch timestamp" that's created (Figure **O**).

In the Arduino IDE, immediately open the Serial Terminal and set the data rate to **9600**. You should be prompted to set the time now. Type in **T** followed by the epoch timestamp you copied (for example, **T1425320521**). Hit Enter and your time should now be set.

Take a look at your clock and confirm that it's telling time as you think it should. Now cover your connections with the heat-shrink tubing and use a heat source to shrink it in place (Figure **P**).

9. PUT IT ALL TOGETHER

Slide the LED assembly into the case, backside first. If it's too tight, you can carefully sand the inside of the case but make sure not to overdo it — you want these parts to fit neatly by friction so you have a nice seamless case. That's it!

Now your clock is assembled and ready for use. Place it on your mantel or your desk and use a wall power supply to keep it running. You'll still need to set the time via USB but once you plug in the wall supply, the clock will keep running after you unplug the USB. Enjoy! ◗

Download the project code and 3D files and share your clock builds at makezine.com/go/3dp-desk-clock.

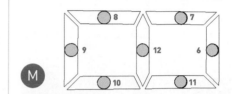

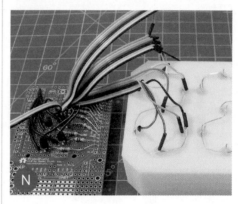

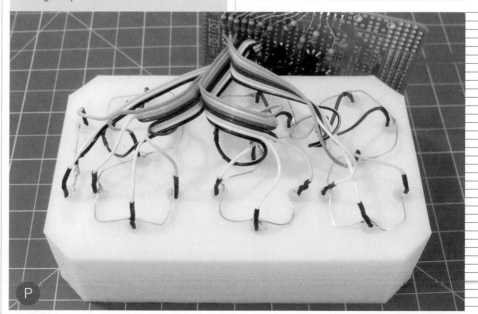

CNC Drip Coffee Stand

Written by Vishal Talwar

VISHAL TALWAR is a software engineer strengthening his grasp on the tougher wares of firmware and hardware as CTO of a lighting startup in Oakland, California. During his spare time, he thinks up ways to clutter his apartment with fun and instructive side projects he designs from scratch.

Time Required: A Weekend
Cost: $40–$60

Materials

- » **Ceramic coffee dripper** I used a Hario V60 02
- » **Walnut plank, ½" thick, 7"×26"** or larger
- » **Cork sheet, ¼" thick**
- » **Wood finish** I used AFM Naturals Oil Wax
- » **Aluminum sheet, 6061 alloy, .040" thick, 5"×5"** or larger, onlinemetals.com
- » **Wood glue** Titebond or similar

Tools

- » **CNC router and bits:**
  - » **Downcut bit, ¼"** for wood and cork
  - » **Straight bit, ¼"** for aluminum
- » **Scissors** or laser cutter
- » **Sandpaper, 150–220 grit**
- » **Flap wheel (optional)**
- » **Router table (optional)** with ⅜" roundover bit

Build a handsome pour-over station to brew your custom cuppa

AFTER SETTING HIS EYES ON AN EXPENSIVE HANDMADE DRIP COFFEE STAND, A FRIEND OF MINE COMMISSIONED ME TO CREATE something similar for his Hario V60 Ceramic Dripper. The overall form is a very simple twist (literally) on other drip stands. I set three objectives for its design: it needed to be easy to clean, use only natural materials (no MDF or plastic), and not appear to be obviously made with CNC.

To these ends, the stand is made of walnut (a good wood for hiding coffee stains) and lined in a couple of places with removable — and washable — cork to absorb unwanted dripping. To add a high-end touch, I CNC-milled a perforated aluminum disk for the base.

1. CUT THE WOOD

Zero the bit on your CNC and make sure the plank you have is wide enough for all your pieces plus another inch or more (at least 7"), in order to maintain a border around all your parts so that the wood doesn't shift or buckle under the force of the cut. I would also recommend screwing the plank into a larger, more easily secured piece of spoilboard.

You can find the DXF vector files, as well as VCarve CAM files embedded with the toolpaths I used for cutting out the various shapes, on the project page online at makezine.com/go/cnc-coffee-stand. The 4 main pieces are the top, 2 identical sides, and the base (Figure Ⓐ).

The top supports the ceramic dripper, so an inner circular profile cut is made to fit the dripper's bottom flange (Figure Ⓑ). An additional pocket cut adds a lip for a cork inlay, which provides a pressure fit for the dripper. To create a snug fit, you'll need to match the radius of the inside hole to the radius of your dripper flange.

Simple rectangular pockets cut into each side fit the top and base, keeping the joints invisible from the outside of the stand. If you don't plan to round over the profile-cut edges later (see Step 4), these rectangles should have overcuts in the corners, known as "dogbones" (Figure Ⓒ), because router bits can't create perfect 90°corners for pocket cuts. A ⅛" bit should do the trick.

2. CUT THE CORK

For the cork inlay in the top, cut a long strip to the circumference of the inner circle, pocket cut it halfway to accommodate the lip, and then squeeze it into the hole (Figure Ⓓ). It is extremely hard to keep cork held down during a cut with a CNC router. Instead, I'd recommend using an ordinary router table for both the profile and the pocket cuts — or better yet, a laser cutter if you have access to one. A sturdy pair of scissors can also get the job done with the help of a template.

The cork for the base is simply a circle cut to fit the pocket, with a little extra room for expansion. If your cork doesn't seem to want to stay flat, steam it for a few minutes and then flatten it under a heavy weight.

3. CUT THE ALUMINUM

Screw the aluminum sheet into a piece of

plywood spoilboard to keep it flat during routing. With a straight ¼" bit, drill the holes — I used Grasshopper 3D software to create and place the perforations in a honeycomb pattern — and then cut out a circle to fit atop the cork in the base (Figure Ⓔ).

After cutting the disk free with shears, I deburred the edges with a flap wheel attached to a drill press. This also gave it a nice brushed look.

4. ROUND OVER THE EDGES (OPTIONAL)

Use a router table equipped with a ⅜" roundover bit to soften the wood's edges and give the piece a more handmade look (Figure Ⓕ). Another benefit of rounding the edges is that you don't need to dogbone the corners of your pocket cuts in the side pieces, since you don't have any 90° angles to account for. You can also use the roundover bit to smooth out the cork lining inside the top piece.

If you prefer a sharper edge to your design, or don't have a router table at your disposal, you can skip this step.

5. FIT AND FINISH

Before gluing the wood, make sure everything fits, including your favorite coffee mug and dripper. Sand the wood and cork to your liking. I used 150 and 220 grit sandpaper in that order until the wood was smooth to the touch.

I applied 2 coats of AFM Naturals Oil Wax to the wood and cork. The oils give a richness and luster to the surface that I prefer to synthetics I've used in the past. Leave oil out of any pockets if you plan to glue it all together.

6. GLUE IT TOGETHER

With the precise cuts of a CNC router, it's possible that your stand fits tightly together using only friction, leaving you the option of disassembling it later. That said, adding a little wood glue will strengthen the design, especially if it's going to hold up to your daily coffee ritual.

Apply glue to just the rectangular pockets in the side pieces and then fit it all together. Wait 10 minutes, then clean off the excess glue with a scraper and a wet rag.

7. GET BREWING

Use your new stand (Figure Ⓖ) to brew up some coffee! After all the work you put in, I guarantee it will be the best coffee you've ever tasted. ⊘

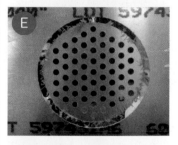

BEFORE
AFTER

Find this project online and download the CNC templates by visiting makezine.com/go/cnc-coffee-stand.

CNC Mechanical Iris

Build this beautiful, eye-opening portal with your CNC or laser cutter.

Written by Chris Schaie

CHRIS SCHAIE
has a background in fine art, graphic design, and advertising. In 2009 he bought his first CNC router and a week later got his first training at "Camp ShopBot" at Maker Faire Bay Area. In 2010 he brought this project to the Faire and won an Editor's Choice blue ribbon. He now owns a small CNC business and teaches CNC at Makerplace in San Diego, California.

Time Required:
6–8 Hours
Cost:
$10–$150

Materials

» **Sheet brass,** .090" thick, 12"×36"
» **Plywood, high quality,** ¾"×24"×48" such as Baltic birch
» **Machine screws, brass, #6 pan head:** 1" (5) and ³⁄₁₆" (30)
» **Washers, Delrin or nylon, #6** (20)
» **Wood screws, brass, #10 pan head** (6)
» **Washers, brass, #10** (10)
» **Fender washer, 1" dia.,** Delrin or nylon
» **Drawer pull, brass**
» **Machine screw, flat head to fit drawer pull,** usually #8 × ¾"
» **Screws (5)** to mount the finished piece to whatever surface you have in mind
» **Thread locking glue** e.g., Loctite

Tools

» **CNC router or laser cutter** I use a ShopBot Buddy 48 with a 4' PowerStick extension. To find a machine or service you can use, visit makezine. com/where-to-get-digital-fabrication-tool-access.
» **Carbide CNC router bits:**
  » **Downspiral bit, ½"** for wood
  » **Endmills, ¼" and ⅛"** for non-ferrous metals
  » **V-bit or engraver**
» **Computer with CAM software** I use Vectric's V-carve Pro.
» **Project files** Download the free DXF drawings and CRV files from makezine.com/go/cnc-mechanical-iris.
» **High-speed rotary tool with cutoff wheel** e.g., a Dremel
» **Drill and drill bits: #36, #27, #18, #9, and countersink** A drill press is highly recommended.
» **Thread tap, #6, and handle**
» **Files or deburring tool**
» **Sandpaper**
» **Double-sided tape**
» **Spindle sander (optional)** makes things much faster and easier

See the iris in action and share your build at makezine.com/go/cnc-mechanical-iris.

THIS ALL STARTED OUT AS A QUEST FOR A BETTER PEEPHOLE FOR MY SHOP DOOR.

I was cruising the interwebs when I found a thread talking about irises. Most of the discussion was debating the pros and cons of a traditional interleaved camera iris in various projects. The biggest problem with that design is that it never entirely closes. Several of us started throwing out ideas for alternatives. *Star Wars* showed up a lot here (the sliding doors in the Death Star, the top hatch on the Millennium Falcon), and when I saw a drawing posted with curved panels that meet in the middle, this project was born.

You can easily cut it on a laser cutter in ¼" acrylic (Figure Ⓐ), plywood, or hardboard. I cut mine from brass and ply on my ShopBot CNC router for about $70 in materials.

1. PREPARE FILES FOR CUTTING

Measure your materials' thicknesses; then, in your CAM software, modify the files as needed to get good cut-throughs and still leave needed tabs. For a laser cutter, change the decorative pocket cuts to an engrave pass, or eliminate them entirely. Output the cutting files (G-code) for your particular machine.

2. CUT THE PARTS

I separated the cut files into groups that used the same bit: ⅛" for cutting out parts, ¼" for pockets, engraver for marking drill holes, and ½" for the plywood backer board.

You'll need a way to hold the material to the bed of the router. If you stick the brass down with double-sided tape, you can use the router to mill holes for hold-down screws; this lets you place them where you won't hit them while cutting.

3. DRILL AND TAP HOLES

Drill holes as indicated on the assembly drawing. Then tap threads in the #36 holes using the #6 tap (a bit of lubricant helps here).

4. CLEAN UP THE PARTS

Here's where the spindle sander helps. Remove any leftover tab material, and use your files or deburring tool to clean up any rough edges.

5. ASSEMBLE

Assemble the actuating outer ring using short #6 screws, then mount it to the plywood by placing wood screws through the open grooves with brass washers behind.

Attach the inner ring, and one corner of each iris leaf, to the plywood with long #6 screws and a Delrin washer between each layer. Connect the 5 linkages using short #6 screws and washers.

For the handle knob, countersink the hole in the drive gear from the back, then use the flat-head screw to secure the drawer pull to the gear. Finally, mount the drive gear with a wood screw and fender washer.

To install the finished unit on a door, I like to rough-cut the hole with a saber saw then use a flush trim bit (with bearing) to rout out the edges.

For exterior use, I made a brass porthole window the size of the aperture and glazed it to weatherproof it. Check my website schaie.com/cnc for a kit. ◗

James Burke

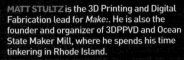

FAB

WRITTEN BY MATT STULTZ

FACTORY

EVERYTHING YOU NEED TO KNOW ABOUT DIGITAL DESKTOP FABRICATION

This is the fourth year of *Make:*'s 3D Printing Shootout, and as we've watched printing mature, we've also observed the emergence of other desktop digital fabrication tools. So this year, along with our printer reviews, we expanded into CNC mills, laser cutters, and vinyl cutters. It's a desktop fabrication revolution!

A few years ago we could have reviewed a handful of CNC machines, but all would have been either DIY builds or largely out of the price range of Makers. Today, the CNC world is quickly expanding its offerings in price, size, and ease of use — finally making them viable tools for the masses.

There's a lot to navigate in this space, and we're here to help you figure out how to get started. These cutting tools should be considered when you want to work with a wider range of materials than 3D printing allows, or need greater precision than hand tools afford. Mills and routers can cut and carve nearly anything, lasers can tightly shape precise designs, and vinyl cutters can go way beyond their name to allow you to create stencils, papercraft projects, and tons of other items. We will always love 3D printers, but expanding the toolchain can only make your projects better.

This issue has something for everyone, from the absolute beginner to the seasoned pro. Now is an exciting time for desktop fabrication. ◔

MEET THE TESTERS

 MATT STULTZ is the 3D Printing and Digital Fabrication lead for *Make:*. He is also the founder and organizer of 3DPPVD and Ocean State Maker Mill, where he spends his time tinkering in Rhode Island.

 KACIE HULTGREN, better known as "Pretty Small Things" in the online 3D printing community, is a multidisciplinary designer focused on set design for live performance. You can find her on twitter: @KacieHultgren

 TOM BURTONWOOD is an assistant professor at the School of the Art Institute of Chicago and is currently working on a 3D printed book project with Chicago cultural historian Tim Samuelson.

 CHRIS YOHE Software developer by day, hardware hacker by night; digital fabrication is his passion. The Makey robots are multiplying way too quickly — send help!

 MATTHEW GRIFFIN is a writer and consultant covering topics such as 3D printing, hobbyist electronics, and more. He has been a regular contributor to *Make:* including the annual 3D printing guides.

 KURT HAMEL is a classically trained mechanical engineer working to bring digital manufacturing and the innovative spirit of the Maker culture to the otherwise conservative shipbuilding industry.

 JASON LOIK is a Rhode Island-based figurative sculptor mainly working in the toy industry. He also spends part of his time teaching at Massachusetts College of Art and Design.

 When **SPENCER ZAWASKY** isn't working for an embedded systems company outside Boston, he's usually 3D printing. Often you'll find him 3D printing at the Ocean State Maker Mill in Pawtucket, Rhode Island.

 SHAWN GRIMES is the Director of Technology at the Digital Harbor Foundation where he works to teach technology and Maker skills to youth and educators.

 CHANDI CAMPBELL builds 3D printers and uses these, the most wonderful of tools, to explore biomimetic structures. Her favorite thing is a good idea and her goal is to generate and/or promote as many as possible.

 LUIS RODRIGUEZ has been tinkering with digital fabrication since 2010. He manages the Maker Studio in Kansas City's science center, Science City, and produces Maker Faire Kansas City.

 JOSH AJIMA is a K-12 3D printing expert and passionate advocate for making in the classroom. He blogs about MakerEd and digital fabrication at DesignMakeTeach.com.

 JIM RODDA, best known to Makers as Zheng3, is an indie video game developer who frequently blogs about 3D printing, Maker culture, and constipated dinosaurs at zheng3.com.

 SAMUEL N. BERNIER is the creative director of Paris design and innovation firm, le FabShop. He's the designer of the articulated Makey, and his first book, *Design for 3D Printing*, is now available from Maker Media.

 CLAUDIA NG is a 3D printing enthusiast who began a small business selling 3D printed plant-toting creatures, Succulent Monsters. A majority of the proceeds are donated to local charities.

 MADELENE STANLEY is a longtime cosplayer who channeled her hobbies into a successful Etsy shop creating 3D printed costume accessories. She currently volunteers for 3D Hubs.

FUSED FILAMENT FABRICATORS

WE PUT THE LATEST MACHINES TO THE TEST TO HELP YOU FIND YOUR NEXT 3D PRINTER

WRITTEN BY MATT STULTZ

"Braq" jointed dragon —
thingiverse.com/make:158409
printed by *Make:* intern Angie Callau.

THERE ARE MORE FILAMENT-BASED DESKTOP 3D PRINTERS AVAILABLE FOR SALE NOW THAN ever, so for this issue we've focused just on the latest releases — either newcomers to the market that we have never tested before, or existing models that have recently been updated enough to warrant retesting. Excitingly, some of these machines are getting the best test results we've ever seen in our reviews.

MACHINE EVOLUTION

The days of 3D printers being made only by tinkerers for tinkerers is long gone. Manufacturers are constantly trying to expand their market by finding new niches. The consumer-focused printer continues to improve with slick metal and injection-molded printer frames taking over where wooden printers once dominated.

CLOSED SYSTEMS

Also notable is the increase in closed filament — either custom spool sizing or requiring chipped spools, which prevents outside vendors' materials from working on the machine. And generally, open source printers seem to be less common right now, especially with so many groups attempting to generate record-setting Kickstarter campaigns with their machines. But some of this year's top-performers remain open.

EASE OF USE

Through this crop of machines, we see a growing trend of easier out-of-box printing experiences. Part of this comes from better bed surfaces, while some is from the increasing number of machines that feature automatic bed leveling — 6 this year. And 5 of the printers use a high-temperature extruder, handy for using exotic filaments like Taulman Bridge or bronzeFill (see page 47).

In short: We all win with these new machines. Now print on. ●

FDM PRINTER PROJECTS:

3D Print Your Head

Use photos to easily create a 3D model of you, then print your noggin for all to admire. A great first project in 3D design and printing. makezine.com/go/print-your-head-in-3d

Finishing 3D-Printed Objects

Friction-weld, rivet, paint, and polish — arm yourself with simple tools and techniques to take your 3D prints to the next level. makezine.com/go/skill-builder-finishing-3dp

Hep Svadja

FDM: HOW WE TEST

HERE'S OUR SCIENTIFIC PROCESS THAT LETS US RANK EACH MACHINE'S CAPABILITIES

WRITTEN BY KACIE HULTGREN

VERTICAL SURFACE FINISH

HORIZONTAL FINISH

DIMENSIONAL ACCURACY

OVERHANGS

BRIDGING

NEGATIVE SPACE

RETRACTION

SUPPORT MATERIAL

Z WOBBLE

TESTING 3D PRINTERS IS ONE OF OUR FAVORITE ACTIVITIES, and each year we strive to make our process more advanced and specific. This year's fused-deposition modeling (FDM; also known as fused filament fabrication or FFF) printer reviews included assessment of 9 test probes that single out each machine's traits, from vertical surface finish to filament retraction.

BIAS-FREE SCORING

To make sure bias did not inadvertently affect scoring, in this year's test we implemented an anonymous review system. Each probe received a unique ID sticker to track the print throughout the review process. Testers logged the probe type, machine name, and print settings when they created any of the probes. At the end of each print, the tester added whether it was successful or the cause of print error. Data validation and comparison between the two forms helped prevent entry errors.

The result was a test probe with a fully recorded provenance that's concealed from view. A print scorer who did not test any of the 3D printers during the event then scored the prints, and a small committee double-checked and confirmed each score. The composite print scores revealed how our test machines stack up (see page 86 for the full machine comparisons).

LOUDNESS TESTING

We also ran a few extra non-scored prints to see how these machines would perform on complex designs users may actually create. The Apollo Astronaut (by Max Grueter) was this year's overnight test print, which showed how each machine would print a multihour design when left largely unattended. And we used the popular print-in-place Articulated Makey (created by tester Samuel Bernier) to check a new parameter this year — loudness. For that, we isolated each printer in a room and used a decibel meter to record the loudest sounds each machine made while producing this design.

Want to test your machine with our probes? Go to makezine.com/go/2015-fdm-test-criteria to download the files, get the review criteria, and share your results. ➋

TAZ 5

THIS OPEN SOURCE WORKHORSE WILL PRINT NEARLY ANYTHING

WRITTEN BY SPENCER ZAWASKY

lulzbot.com

MANUFACTURER
LulzBot

PRICE AS TESTED
$2,200

BUILD VOLUME
298×275×250mm

BED STYLE
Heated glass covered
with PEI print surface

FILAMENT SIZE
3mm

OPEN FILAMENT?
Yes

TEMPERATURE CONTROL?
Yes, tool head (300°C max)

PRINT UNTETHERED?
Yes (SD card)

ONBOARD CONTROLS?
Yes, LCD screen

HOST/SLICER SOFTWARE
Cura LulzBot Edition
(Slic3r still fully supported)

OS
GNU/Linux, Mac, Windows

FIRMWARE
LulzBot Taz 5 Single extruder
(Open Marlin Based)

OPEN SOFTWARE?
Yes, Respects Your Freedom

OPEN HARDWARE?
Yes, Respects Your Freedom

MAXIMUM DECIBELS
68.4

Available at
Maker Shed

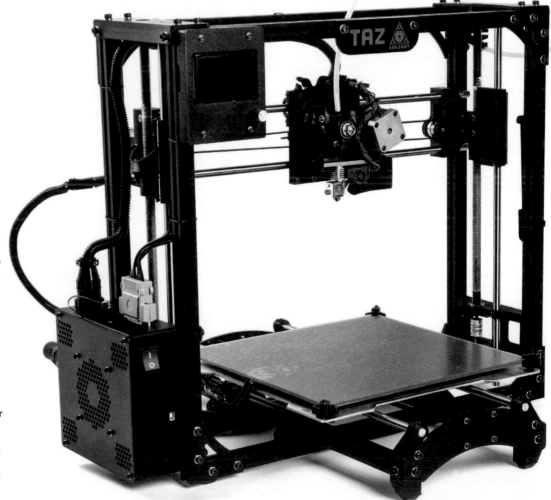

Gunther Kirsch

AT A GLANCE YOU CAN BE FORGIVEN IF YOU DON'T NOTICE ANY DIFFERENCE BETWEEN THE TAZ 4 AND THE TAZ 5. But if you've been following LulzBot and their meteoric rise this past year, you'd have noticed that their smaller offering, the LulzBot Mini (see page 34) featured a new "Hexagon" all-metal hot end, a self-calibrating platform, a frosted orangey-brown Polyetherimide (PEI) print surface, and Cura support. The Taz 5 brings three out of four (no self-leveling bed) of the Mini's advances back to LulzBot's flagship design, and the results reflect that.

METAL PROVES ITS METTLE

The new hot end includes an additional blower aimed straight at the heat-break. Being all-metal, it is rated to 300°C (up from 240°C), expanding the variety of materials to include various nylons, polycarbonate, and PETT.

The hot ends now ship with 0.50mm nozzle apertures. The larger opening makes it easier to put out more plastic per layer — and thus take advantage of the nearly cubic foot of print volume more quickly. It also reduces the stress of printing the newer filaments that contain various powdered materials, such as bronze, copper, and steel.

Nylon? No problem. A little glue stick, download the profile from lulzbot. com, slice, and print. The firmware even includes a dozen or so presets for various filaments. This is the sort of experience that makes a user eager to try interesting new materials.

However, that new capability comes with the arguable loss of some fine detail, though you can get the 0.35mm nozzle if you really want it. (In fact, the old hot end is still available if you want that too.) And the Taz retains its easy-to-swap tool head design — undo one screw, unplug a few cables, and you're done. Although, depending on what you swap, you may have to change firmware to match.

EASY ON, EASY OFF

To the touch, the PEI seems like nothing special, a bit like an acid-etched glass surface. But when the plastic hits it, the magic happens.

ABS and HIPS both stick admirably to the PEI around the recommended 110°C range, and PLA does even better. But what is delightfully surprising is that, back down around room temperature, HIPS and ABS pop right off.

For materials that don't play as nicely with PEI, particularly nylons, some glue stick comes to the rescue. The Taz's industry-leading documentation offers a table of materials and temperatures (nozzle, bed, and removal), as well as whether to use a glue stick.

A CURA FOR WHAT AILS

Along with the Mini, the Taz 5 (and Taz 4) are now supported in Cura. LulzBot provides you with its own branded version of Cura, though it's not strictly necessary.

We tested using the default settings and got great results with super-fast slicing. LulzBot has created six profiles (fine, medium, and fast, for 0.5mm or 0.35mm nozzles) for each printer and the acceptable materials, as well as a full array of Slic3r profiles.

The stellar documentation has been augmented to include a full walk-through for Cura, as well as additional supporting information for using various materials.

CONCLUSION

It was hard to improve on the Taz 4, but a new hot end, a durable sticky-when-hot print surface, and some added, easy-to-use software options have certainly done it. Even better, the Taz 5 brings no controversial or ruinous downgrades to a successful design. And while some may lament the lack of an automated platform-level calibration, others may look at that as an unnecessary gimmick. The Taz 5 is a worthy successor, still true to its open heritage, and a match for any challenger. ◐

> When the plastic hits the print surface, the magic happens

PRINT SCORES

0 1 2 3 4 5

- Vertical Surface Finish
- Horizontal Finish
- Dimensional Accuracy
- Overhangs
- Bridging
- Negative Space
- Retraction
- Support Material
- Z Wobble

TOTAL

35

PRO TIPS

Muster your patience. That beautiful, fresh print will come off 100 times more easily if you let the platform cool down.

WHY TO BUY

Do you want big beautiful prints and a machine that is not only ready to take on anything but also ready to upgrade into the future? Then the Taz 5 is for you. The new all-metal hot end combined with the heated bed will take on any filament you throw at it. Open source everything means that when LulzBot or the community make updates, you can incorporate them as you choose.

RESULTS

PRINTRBOT PLAY

Available at **Maker Shed**

HIGH QUALITY AND LOW COST COMBINE TO MAKE THIS A GREAT ENTRY-LEVEL PRINTER

WRITTEN BY TOM BURTONWOOD

printrbot.com

MANUFACTURER
Printrbot

PRICE AS TESTED
$399

BUILD VOLUME
100×100×130mm

BED STYLE
Unheated aluminum

FILAMENT SIZE
1.75mm

OPEN FILAMENT?
Yes

TEMPERATURE CONTROL?
No

PRINT UNTETHERED?
Yes (SD card)

ONBOARD CONTROLS?
No

HOST/SLICER SOFTWARE
Cura

OS
Linux, Mac, Windows

FIRMWARE
Marlin

OPEN SOFTWARE?
Yes, Cura/CuraEngine : AGPLv3

OPEN HARDWARE?
Yes, CC BY SA

MAXIMUM DECIBELS
82.3

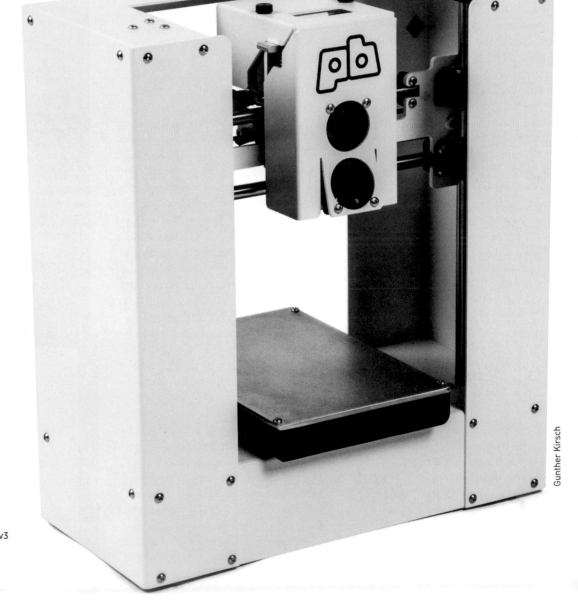

Gunther Kirsch

BROOK DRUMM AND PRINTRBOT ARE AN INSPIRATION TO MAKERS AND HARDWARE HACKERS EVERYWHERE. Their company credo seems to be a healthy combination of minimalism and irreverence from which a good number of innovative, thought-provoking products have emerged. The Printrbot Play is another excellent 3D printer from this lineage in terms of print quality, speed, and form factor. While the design is clearly aimed at an educational audience of young Makers and their teachers, the Play should appeal to anyone on a budget looking for an entry-level printer.

STURDY WITHOUT BEING TOO HEAVY

While it shares some obvious comparisons to the metal Plus released earlier in the year, the Play seems to be the first generation of Printrbot machines designed from the ground up with a metal frame in mind. Constructed from powder-coated aluminum and steel, the Play is rigid and sturdy without being too heavy. The bottom of the frame is open, allowing for easy access to motors and electronics, and four rubber stoppers keep the Play from wandering across the desk, a problem I have encountered with the Simple.

Other improvements include a much more accessible microSD card slot and a top-mounted filament bracket that keeps the overall footprint small. However, one improvement sorely missing from pretty much every Printrbot to date is an off switch. This might be a petty request, but it would be really great to see something like this added in the future.

MODEST BUILD SIZE

The hot end and nozzle are hidden behind a handy shroud that also holds both the extruder and cooling fans in place. The shaped metal frame serves as an enclosure of sorts, keeping curious fingers away from the majority of the spinning, heating and moving parts. On the flip side, if I needed to replace the X-axis stepper I would have to disassemble the entire frame just to reach the screws holding the motor in place.

The Printrbot Play build size is not huge, clocking in at a fairly modest 4×4×5", good for school projects and prototyping small assemblies but not great for any medium-sized pieces and above. Despite being a metal box, the Play is surprisingly lightweight and would be an ideal machine for taking on the road for a local print demo.

QUICK, HIGH-QUALITY PRINTS

Although it looks a little sluggish, the Play is actually very brisk — zipping prints out quickly without significantly sacrificing quality. On the XY test probe it stood head and shoulders above the other machines, showing no signs of echoes or resonance on the X- or Y-axis. The horizontal surfacing probe was also good with only a small blemish on the top of the sphere. Overhang and bridging probes both showed minor problems with sagging plastic. Users should probably either print their own cooling fan ducting or give some thought to optimum rotation and/or placement of models to get the most out of the cooling fan. In addition to print quality, speed, and form factor, we also tested how loud each printer was during operation. The Play was one of the louder machines we tested, but I didn't find the familiar warbling to be super distracting.

CONCLUSION

Like the Simple and Plus before it, the Printrbot Play is an affordable and solid platform, ideal for educators and beginners alike. While it has some shortcomings in terms of noise and direct cooling, the test-print quality held up very well in comparision with those of the other machines that cost considerably more. Printrbot has developed a made-in-the-USA business model that keeps the prices of their products low but the quality high, and they continue to flip the para-digm of doing something both innovative and useful. ◐

> Printrbot has developed a business model that keeps the prices of their products low, quality high

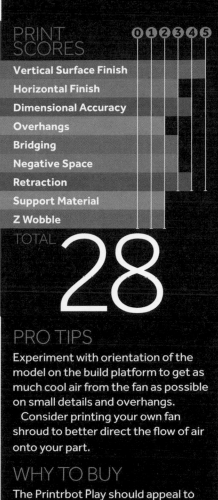

PRINT SCORES

0 1 2 3 4 5

- Vertical Surface Finish
- Horizontal Finish
- Dimensional Accuracy
- Overhangs
- Bridging
- Negative Space
- Retraction
- Support Material
- Z Wobble

TOTAL
28

PRO TIPS

Experiment with orientation of the model on the build platform to get as much cool air from the fan as possible on small details and overhangs.

Consider printing your own fan shroud to better direct the flow of air onto your part.

WHY TO BUY

The Printrbot Play should appeal to anyone on a budget looking for an entry-level 3D printer.

RESULTS

ULTIMAKER 2 EXTENDED, ULTIMAKER 2 GO

Available at **Maker Shed**

THEY'RE LIKE CLONES OF THE ULTIMAKER 2, BUT IN A GOOD WAY

WRITTEN BY KURT HAMEL

ultimaker.com

Kelly Egan

MANUFACTURER Ultimaker

PRICE AS TESTED $ 2,788 (Extended) | $1,335 (Go)

BUILD VOLUME 223×223×305mm (Extended) | 120×120×115mm (Go)

BED Heated Glass (Extended) | Unheated glass (Go)

FILAMENT SIZE 3mm

OPEN FILAMENT? Yes

TEMPERATURE CONTROL? Yes, tool head (260°C max)

PRINT UNTETHERED? Yes, SD card reader

ONBOARD CONTROLS? Display screen and control wheel

HOST/SLICER SOFTWARE Cura

OS Mac, Windows

FIRMWARE Marlin

OPEN SOFTWARE? CuraEngine is released under terms of the AGPLv3 License.

OPEN HARDWARE? Yes, hardware is CC BY NC

MAXIMUM DECIBELS 77.5 (Extended) 75.2 (Go)

AT FIRST GLANCE, THE ULTIMAKER 2 EXTENDED MAY SEEM LIKE JUST A BIGGER, more expensive version of the much-heralded Ultimaker 2, and the Ultimaker 2 Go may seem like just a smaller, less expensive version. And that's more or less accurate — with one small, cost-saving exception: The Go is not equipped with a heated build plate.

LET'S GO!
Setup for both was easy; a simple and almost unnecessary one-page startup guide got the printer turned on in less than two minutes. The display screens guided me through loading material and bed leveling (which is manual) without any issues. A preloaded 4GB SD card positioned me to make my first print minutes after opening the box. My only gripe up to this point was a stiff selector wheel on the Extended (compared to a much smoother one on the Go). Maybe it just needs to be broken in?

Cura is probably the easiest and fastest slicing software I've encountered. There's not much to say there. However, while Cura offers a lot of control, there are some settings that need to be done on the machine, like temperature. It would be better to do everything in the slicing software so there's no fumbling around at the machine.

FINE TUNING
During testing, the Extended printed away without issue while the Go had some intermittent extrusion anomalies that caused some porosity in some of the test prints. It seemed like the Go's Bowden extruder was trying to move more filament than the hot end could melt, which would induce some slipping at the drive gear. This was hard to explain because the Extended — which was equipped with identical parts — did not seem to have that issue. Fortunately for the Go, this porosity did not appear in any of the tests designed to look for it.

Changing some feed rates and temperature settings made this problem vanish, but I was disappointed that I needed to override default settings to get the best print. A novice user might operate it with less than optimal settings just because they don't know any better.

That said, both machines produced extremely nice test prints, even without my intervention.

PACKAGING WORTH SAVING
Unboxing these printers was a pleasant surprise. Where I thought I would encounter cheap, flaky foam inserts I found instead high-quality reusable foam shells that double as a carrying case, held together by durable nylon straps for lifting and transport. The foam shell protecting the Go was especially fancy, and seemed like it also would function as a printer stand.

CONCLUSION
You don't need to look far to hear accolades for the original Ultimaker 2, and I can't say I disagree with them — it's a top-caliber machine that gets great results. The Extended and the Go have little reason to perform any differently, so if you're a fan of the Ultimaker 2 you will be a fan of the entire family. These machines are great for everyone who can afford them — Makers, artists, teachers, students, and engineers alike.

Ultimaker is producing three essentially identical machines in different sizes rather than mixing features and build volumes to come up with three different price points. That's a good thing: It makes the decision of which one to get less complicated. Plus, there's merit to interchangeability. Not only does it make sourcing parts easier, it also assists with seeking help. For instance, the profile settings I gleaned from the internet to tune the Go ended up producing better work on the Extended as well. So, if you're going to expand your product line by cloning an existing product, you might as well clone one that's getting so much applause. ◉

> If you're going to expand your product line by cloning an existing product, you might as well clone one that's getting so much applause

ULTIMAKER 2 EXTENDED PRINT SCORES

	0	1	2	3	4	5
Vertical Surface Finish						
Horizontal Finish						
Dimensional Accuracy						
Overhangs						
Bridging						
Negative Space						
Retraction						
Support Material						
Z Wobble						

TOTAL 30

ULTIMAKER 2 GO PRINT SCORES

	0	1	2	3	4	5
Vertical Surface Finish						
Horizontal Finish						
Dimensional Accuracy						
Overhangs						
Bridging						
Negative Space						
Retraction						
Support Material						
Z Wobble						

TOTAL 30

PRO TIPS
A known issue with these machines is that certain print profiles can cause the extruder to outpace the hot end, which causes the extruder to slip, which makes the print porous. It helped to increase the default extrusion temperature from 210°F to 230°F. Other tweaks can be made in Cura and are easy to find with an internet search.

WHY TO BUY
If you're a fan of the sleek look and great prints you get with the Ultimaker 2, but need a different form factor (bigger or smaller), the Extended and Go are perfect options.

RESULTS

ULTIMAKER 2 EXTENDED

ULTIMAKER 2 GO

LULZBOT MINI

THIS SMALL MACHINE WITH A CLEVER AUTO-LEVELER IS GREAT FOR BOTH BEGINNERS AND EXPERTS WRITTEN BY JIM RODDA

THIS PLEASANTLY INDUSTRIAL LITTLE BOT WOULD BE AT HOME IN ANY WORKSHOP, CLASSROOM, OR MAKERSPACE. The frame is sturdy and built from matte black sheet metal. Internal cables are nicely contained and tied down but still accessible.

Cura LulzBot Edition's default settings produce attractive, functional prints with reasonable speed. Advanced users will want to dig into the expert settings posthaste.

HANDS-FREE BED LEVELING

Watching the LulzBot Mini's auto-leveling feature is a joy for anyone who's spent an afternoon twiddling thumbscrews in pursuit of a level bed. It touches the nozzle to contact points at the bed's corners, and compensates virtually. The corners of the bed are not physically adjusted.

The Mini produced test print after test print with no manual tweaking or recalibration, and did not crash, jam, or otherwise fail during the test period. Prints stick fast to the heated bed during printing and are easy to remove once the bed has cooled.

NOT TOO HIP ON HIPS

LulzBot has focused in on HIPS as their filament of choice. We ran into uncharacteristic issues with both LulzBots during testing; rerunning them in PLA resulted in excellent prints by the Mini, jumping the overall score from 22 to 32.

Oddly for a portable device, the printer lacks Wi-Fi or SD card support, and must be physically tethered via USB to an external computing device in order to function. The Mini doesn't have any external display devices to indicate print progress or error codes.

CONCLUSION

With only a few minor downsides, the LulzBot Mini is a perfect match for high school and university STEM labs, Makerspaces, basement tinkerers, or anyone looking for a sturdy, reliable, portable bot at a reasonable price point. This machine is a good choice as a first exposure to 3D printing or as a small-scale workhorse for an expert user. ◗

> If you've been 3D printing for a while, the auto-leveling feature is mind-blowing the first time you see it

PRINT SCORES — 0 1 2 3 4 5

Vertical Surface Finish
Horizontal Finish
Dimensional Accuracy
Overhangs
Bridging
Negative Space
Retraction
Support Material
Z Wobble

TOTAL 32

MANUFACTURER LulzBot
PRICE AS TESTED $1,350
BUILD VOLUME 152×152×158mm
BED STYLE Heated glass covered with PEI print surface
FILAMENT SIZE 3mm
OPEN FILAMENT? Yes
TEMPERATURE CONTROL? Yes, Tool Head (300°C max)
PRINT UNTETHERED? No
ONBOARD CONTROLS? No. Power switch only.
HOST/SLICER SOFTWARE Cura LulzBot Edition
OS Debian, Mac, Windows
FIRMWARE Open, Marlin
OPEN SOFTWARE? Yes. Cura LulzBot Edition is derived from Open Source Cura
OPEN HARDWARE? Yes. GPLv3 and/or CC BY SA 4.0
MAXIMUM DECIBELS 86

PRO TIPS

Use a dedicated PC, tablet, or Raspberry Pi to run long prints while keeping your main workstation untethered.

Wait for prints to cool and they'll pop off the print bed like magic.

Grab a roll of 3mm PLA filament with your purchase, as the printer only ships with a short HIPS sample.

WHY TO BUY

Easy to use and bundled with outstanding documentation, the LulzBot Mini reliably produces excellent prints at a reasonable price. Useful as a first exposure to 3D printing or as a small workhorse for an expert, and in Makerspaces or high school or university STEM labs.

RESULTS

lulzbot.com

Gunther Kirsch

CHOOSE FREEDOM.

ZORTRAX M200

Available at
Maker Shed

A GREAT-LOOKING MACHINE, WITH PRINT QUALITY TO MATCH
WRITTEN BY SPENCER ZAWASKY

THIS PRINTER IS ABOUT PRINTING, NOT FIDDLING. And print it does — with distinction. While the retraction performance and bridging could use some improvement, they are good nonetheless, and the surface quality and (lack of) Z-wobble are really something to see.

Visually, it is stark and stylish: A featureless black frame with a cold blue fluorescent display echoes the cool white chamber lights, which are dimmed by tinted acrylic panels. It is, in a word, sleek.

SIMPLICITY, FOR BETTER OR WORSE
The firmware and software are correspondingly simple. The bright, legible display has a small handful of entries each with a smaller handful of items. It does what it needs to do and nothing more.

The M200 works only with the Z-Suite software, and the Z-Suite software will not download without a Zortrax serial number. It will not install without a Zortrax serial number. And, strictly speaking, the required proprietary software provides profiles for only Zortrax filament.

STRANGE BUT NECESSARY
The perforated build platform is removable, though mounted firmly by pegs and magnets to the Z-stage. Electrical connections are by dual cable for heat and temperature sensing.

While these connectors are perfectly suited for the occasional connect and disconnect, they seem ill chosen for the structured world of Zortrax, where you must wait for the platform to cool (or use an oven mitt), power off, unplug, and remove the platform to remove each print.

The Z-Suite software always prints a raft and that raft always adheres to the surface. Removing the raft from the platform involves a metal blade and a very special blend of finesse and violence. Thoughtfully, a fine specimen of such an implement is provided, along with a dozen or so other helpful accessories, from work gloves to tiny nozzle-unclogging spikes.

CONCLUSION
Print after print after print, the M200 pumped out consistently excellent results. If you really just want to print well and you're willing to pay a bit more for looks, forgive a few small blemishes, have no interest in upgrades, rebuilds, or tweaks, and are happy to buy the filament from the manufacturer, the M200 is the printer for you. ◈

It is, in a word, sleek

MANUFACTURER Zortrax
PRICE AS TESTED $2,000
BUILD VOLUME 200×200×180mm
BED STYLE Heated PCB perf board
FILAMENT SIZE 1.75mm
OPEN FILAMENT? No
(proprietary slicer only supports their filaments)
TEMPERATURE CONTROL? No
PRINT UNTETHERED? Yes (SD card)
ONBOARD CONTROLS? Yes
HOST/SLICER SOFTWARE Z-Suite
(vendor proprietary, Serial No. required)
OS Mac (requires Mono SDK), Windows
FIRMWARE Closed
OPEN SOFTWARE? No
OPEN HARDWARE? No
MAXIMUM DECIBELS 64.5

zortrax.com

PRO TIPS
Impatient users will want to keep an oven mitt handy — it can take a while for the platform to cool enough to remove it to liberate your prints.

The firmware has no "cancel" feature, but powering off works well enough.

WHY TO BUY
You pay a bit more for the M200, but in return can focus on enjoying the results rather than tuning the process.

Gunther Kirsch

RESULTS

PRINT SCORES

	0	1	2	3	4	5
Vertical Surface Finish						
Horizontal Finish						
Dimensional Accuracy						
Overhangs						
Bridging						
Negative Space						
Retraction						
Support Material						
Z Wobble						

TOTAL
28

MANUFACTURER Dremel
PRICE AS TESTED $999
BUILD VOLUME 230×150×140mm
BED STYLE Unheated acrylic with BuildTak
FILAMENT SIZE 1.75mm
OPEN FILAMENT? No (Voids warranty)
TEMPERATURE CONTROL? Yes, tool head (230°C max)
PRINT UNTETHERED? Yes (SD card and touchscreen controls)
ONBOARD CONTROLS? Yes
HOST/SLICER SOFTWARE Dremel3D
OS Mac, Ubuntu, Windows
FIRMWARE Closed
OPEN SOFTWARE? No
OPEN HARDWARE? No
MAXIMUM DECIBELS 80.4

3dprinter.dremel.com

PRO TIPS

Use Autodesk Meshmixer (which includes a profile for the Idea Builder) to add support. Simplify3D is another alternative.

Get yourself a set of feeler gauges for when you lose that leveling card.

Set the bed screws tight enough so you can feel the nozzle scrape the leveling card, but loose enough that it still moves freely.

WHY TO BUY

This fully enclosed printer is easy to use, reliable, and produces decent prints. The simple software, touchscreen interface and warranty make it a good value for beqinners or for those who don't want to play with hardware.

RESULTS

Gunther Kirsch

3D IDEA BUILDER

Available at
Maker Shed

A BASIC, RELIABLE PRINTER WITH A WARRANTY
WRITTEN BY SHAWN GRIMES

THIS FULLY ENCLOSED, EASY TO USE, RELIABLE MACHINE COULD OPEN UP 3D printing beyond the realm of early adopters. Though the warranty and closed-source nature can be a bit restrictive, the Idea Builder (aka 3D20) provides quality prints.

SIMPLY SMART
The printer ships with two sheets of Dremel-branded BuildTak (to keep prints from sticking to the bed), a leveling card, a plastic scraper, and a nozzle de-clogging tool. Its easy-to-follow videos can get you up and printing in about 30 minutes.

The addition of the BuildTak and leveling card are two signs that Dremel is trying to offer an all-inclusive printing ecosystem. The leveling card is such a seemingly simple addition, taking the guesswork out of what paper to use for manual bed leveling.

SOFTWARE FLAWS, PROPRIETARY PLASTIC
The Dremel 3D software is easy to use but is lacking in some pretty fundamental features — most notably the ability to add support material directly. Dremel instead wants you to use Autodesk Meshmixer, which could be a challenge to someone new to 3D printing. This missing feature may stem from Slic3r v1.0.1 being their base slicer — that version of the software had some issues with adding support that could be easily removed. Fortunately, this is a software issue that could be readily resolved in future updates.

Dremel also requires proprietary filament. I understand that if you offer a warranty, you want to ensure users aren't using impure filament, but with only 10 colors available — and most of them translucent — your options are limited. Additionally, at $30 for only 0.5 Kg, the costs can add up quickly.

CONCLUSION
The Idea Builder is poised to attract those looking for stable technology with its reliable printing, ease of use, and decent quality. Schools and users who want a dependable, affordable tool that comes with a warranty will enjoy this machine. ⊘

Dremel is trying to offer an all-inclusive printing ecosystem

PRINTRBOT SIMPLE

THIS SOLID MACHINE IS DESIGNED FOR EASE OF USE, RELIABILITY, AND LOW PRICE WRITTEN BY SAMUEL N. BERNIER

IT'S BEEN A PLEASURE WATCHING PRINTRBOT EVOLVE SINCE THEIR 2011 KICKSTARTER CAMPAIGN, offering affordable and easy 3D printer kits. Like many brands, they used laser-cut wood in their early models, then moved to more reliable, scalable metal. In doing so, they continue to keep the prices low and the quality high, as seen in the Simple.

NEW EXTRUDER, Z-LEVELING TOOL

The Simple has received some cool upgrades such as a new machined aluminum extruder, which gives more leverage for easier loading and better tolerances, but also enables better flexible filament extrusion. The metal print bed has also been updated to thick, unpainted aluminum, with holes for accessibility.

The Z-leveling probe is a surprising tool in a $599 machine, and it provides straightforward printing without further bed adjustment.

LESS IS MORE

If you want a heated bed or an LCD interface, you will have to buy them separately. The Simple has no on/off button, no colorful lights, not even Wi-Fi or internal memory — and this is why people love it so much. Let's not forget that 3D printers are tools, not toys, and more features mean more things can go wrong.

WORKS AS RECOMMENDED

We used Cura, with the Pronterface plugin and imported Printrbot presets available on their website, which gave us immediate results. The provided settings worked fine as long we used Printrbot's filaments, but we experienced constant clogging with PLA from other brands, though adjusting the temperature and fan helped.

The microSD card slot is still a challenge to use, which is why most users will prefer using the classic USB connection.

CONCLUSION

The Simple is perfect for anybody getting started with personal 3D printing. There are plenty of videos and documents to help you, and the machine can be easily fixed and upgraded as needed. It is also one of the best kits available and can be assembled in a single afternoon. ⊘

With the Simple, Printrbot continues to keep the prices low and the quality high

PRINT SCORES

	0	1	2	3	4	5
Vertical Surface Finish						
Horizontal Finish						
Dimensional Accuracy						
Overhangs						
Bridging						
Negative Space						
Retraction						
Support Material						
Z Wobble						

TOTAL 28

MANUFACTURER Printrbot
PRICE AS TESTED $599
BUILD VOLUME 150×150×150mm
BED STYLE Unheated aluminum ($150 heated upgrade available)
FILAMENT SIZE 1.75mm
OPEN FILAMENT? Yes
TEMPERATURE CONTROL? Yes, Bed (80°C max)
PRINT UNTETHERED? Yes (MicroSD card)
ONBOARD CONTROLS? No, but LCD add-on available
HOST/SLICER SOFTWARE Cura with Pronterface UI
OS Linux, Mac, Windows
FIRMWARE Marlin
OPEN SOFTWARE? Cura/CuraEngine : AGPLv3
OPEN HARDWARE? Auxiliary design files: CC BY-NC-SA 3.0
MAXIMUM DECIBELS 79.1

PRO TIPS

When using the optional metal spool rack with spools from other brands, make sure they don't obstruct the Z-axis when the extruder goes up.
Use painter's tape on the aluminum bed when using PLA.
For easier extruder loading, cut the filament at 45°.

WHY TO BUY

The Printrbot Simple is affordable, easy to use, very well documented, and most of all, it works.

printrbot.com

Gunther Kirsch

RESULTS

PRINT SCORES

	0	1	2	3	4	5
Vertical Surface Finish						
Horizontal Finish						
Dimensional Accuracy						
Overhangs						
Bridging						
Negative Space						
Retraction						
Support Material						
Z Wobble						

TOTAL
30

MANUFACTURER Deezmaker

PRICE AS TESTED $849 ($999 Limited Edition)

BUILD VOLUME 125×150×125mm

BED STYLE Unheated acrylic

FILAMENT SIZE 1.75mm

OPEN FILAMENT? Yes

TEMPERATURE CONTROL? Yes, tool head (295°C max)

PRINT UNTETHERED? Yes (SD card)

ONBOARD CONTROLS? Yes (Autostart from SD card, manual reset)

HOST/SLICER SOFTWARE Repetier (host), Cura (slicing)

OS Linux, Mac, Windows

FIRMWARE Open, Custom Azteeg x2D Firmware

OPEN SOFTWARE? Yes, Cura AGPLv3 (slicing); No, Repetier Is Closed Source (host)

OPEN HARDWARE? No

MAXIMUM DECIBELS 56

PRO TIPS

Go online for the latest setup information.

Take your favorite (or current) print, throw it on the SD card as auto0.g and hit reset. Voila: Auto printing and your computer is freed for other uses.

Take your time to get the build plate leveled in. Once it was leveled it printed rock solid.

WHY TO BUY

Small form factor, portability, and open design come together to deliver much better than expected results, at a slightly higher than expected price.

deezmaker.com

Gunther Kirsch

RESULTS

BUKITO

THIS SMALL BUT MIGHTY MACHINE IS PLANE, TRAIN, AND AUTOMOBILE COMPATIBLE WRITTEN BY CHRIS YOHE

IF YOU GO BY ITS STATURE, YOU MIGHT GET THE NOTION THAT THIS SMALL PRINTER is just another mini with middling results. You would be wrong. The Bukito scored well in our tests and seems to be as portable as Deezmaker claims. The primary drawback is a small build volume.

LEAN, MEAN PRINTING MACHINE

The first thing that catches your eye is the somewhat unique design. The open arms, rails, and mechanics are accentuated with laser cut and engraved housings around some of the stepper motors and electronics. The standard belts were replaced with a synchromesh drive system and the extruder is safely tucked behind a metal housing. A keyed power connector makes wiring it up a snap, and when we plugged it in the printer hummed to life, printing out default print-in-place spinning gears — a pleasant surprise.

TO-GO BOX, PLEASE

We didn't have enough time to fully test one of the most touted features: portability. But moving the machine from table to table — even during printing — elicited no ill effects, and the sturdy nature of the machine made us feel that minor disturbances wouldn't affect the prints too much.

According to the manufacturer, the machine works great in a variety of environments, requiring only a 12V 5Amp power source, making it plane, train, automobile, and apparently drone approved. (Use your own discretion in choice of venue.)

CONCLUSION

What seemed to be another cute, small, portable machine provided great results. While it lacks the polished, boxy finish of some of the prosumer machines, the unique look and cheeky accents add to its charm and spunky nature. ●

BUKOBOT

In addition to the Bukito, we got a chance to play with a not-quite-ready-for-review, next-generation Bukobot. The team is working hard to upgrade the hardware and has some great things on the way for this open source machine. Stay tuned next year to see what Deezmaker has lined up.

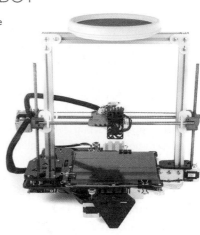

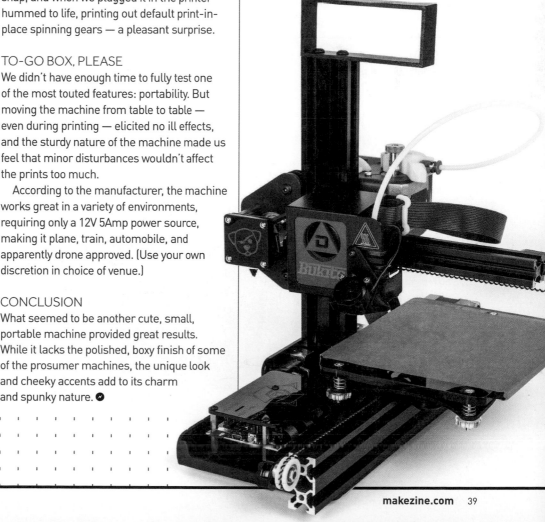

DA VINCI 1.0 JR.

A BEGINNER PRINTER FOR THOSE WHO DON'T NEED HIGH-QUALITY PRINTS WRITTEN BY JASON LOIK

PRINT SCORES 0 1 2 3 4 5

Vertical Surface Finish
Horizontal Finish
Dimensional Accuracy
Overhangs
Bridging
Negative Space
Retraction
Support Material
Z Wobble

TOTAL **24**

MANUFACTURER XYZprinting
PRICE AS TESTED $349
BUILD VOLUME 150×150×150mm
BED STYLE Unheated glass with custom tape
FILAMENT SIZE 1.75mm
OPEN FILAMENT? No (Chipped)
TEMPERATURE CONTROL? No
PRINT UNTETHERED? Yes (SD card)
ONBOARD CONTROLS? Yes
HOST/SLICER SOFTWARE XYZware
OS Mac, Windows
FIRMWARE Closed
OPEN SOFTWARE? No
OPEN HARDWARE? No
MAXIMUM DECIBELS 61.9

us.xyzprinting.com

THE CHEAPEST ADDITION TO THE XYZ FAMILY IS THE DA VINCI 1.0 JR., ONE OF THE LEAST EXPENSIVE MODELS WE TESTED this year. For $350, you get user-friendly software, enclosed printing, and an online community.

The da Vinci Jr. is generally a simple, easy-to-use machine — it is calibration free, so needs no painful bed leveling, and offers a decent print volume, 5.9" on all axes. However, its use of proprietary chipped filament will push your cost up, and adds an unnecessary level of complexity when you go to change the spool. And it would have been nice to include an intro to 3D printing pamphlet, DVD, USB stick, carrier pigeon, or anything to help ease an enthusiast over the common pitfalls of this nuanced game.

QUALITY ISSUES

The good news: It only costs $350. The bad news: It prints like it only costs $350. The results are underwhelming. One notable struggle: None of the testers were able to complete our overnight astronaut print successfully, instead receiving a plate of spaghetti every time it was run (see bottom right). The hardware is an interesting mix of components. (It even looks as though they are using a few inkjet printer parts.)

But the guts seem fairly sturdy, so most likely the lack of print quality is a result of software. While it's fairly user friendly, the slicing is slower than Cura and seems to have trouble handling anything less than a perfect model. Hopefully XYZ will update soon.

CONCLUSION

The da Vinci Jr. is made to look pretty on the store shelves — maybe catching the eye of a parent that didn't know 3D printers were getting so affordable. This machine is for a consumer that is interested in running an occasional novelty print, where the expectation of quality isn't all that high. Be sure to consider the long-term expense of higher-priced chipped filament. ◉

> The good news: It only costs $350. The bad news: It prints like it only costs $350

PRO TIPS

The machine only comes with one of its proprietary spool holders and new filament doesn't ship with replacements. Do yourself a favor and order extras to avoid having to go through a needless ritual every time you want to switch out a color.

WHY TO BUY

Appropriate (and appropriately safe) for families with small children, the da Vinci Jr. brings 3D printers into the realm of the impulse buy.

Matt Stultz

RESULTS

PRINT SCORES 0 1 2 3 4 5

Vertical Surface Finish	
Horizontal Finish	
Dimensional Accuracy	
Overhangs	
Bridging	
Negative Space	
Retraction	
Support Material	
Z Wobble	

TOTAL 33

MANUFACTURER SeeMeCNC
PRICE AS TESTED $999
BUILD VOLUME 280mm diameter×375mm
BED STYLE Heated glass
FILAMENT SIZE 1.75mm
OPEN FILAMENT? Yes
TEMPERATURE CONTROL? Yes, tool head (240°C max)
PRINT UNTETHERED? Yes (SD card)
ONBOARD CONTROLS? Yes, scroll knob and LCD
HOST/SLICER SOFTWARE MatterControl
OS Linux, Mac, Windows
FIRMWARE Repetier
OPEN SOFTWARE? Yes
OPEN HARDWARE? Yes
MAXIMUM DECIBELS 61.2

PRO TIPS

There are a few needed items for assembly that don't come with this kit; download the manual ahead of time so you can get these items before the kit arrives.

The build will require some soldering so be sure to have a decent iron at the ready.

Read through all of the instructions before beginning, it will save you time later.

WHY TO BUY

With a super large build volume, onboard controls, and a reasonable price tag, this printer offers a great value for those looking for a full-featured machine.

RESULTS

ROSTOCK MAX V2

THIS 4' TALL, FULL-FEATURED KIT MACHINE IS GREAT FOR EXPERIENCED USERS WRITTEN BY MATT STULTZ

THIS KIT MACHINE OFFERS HUGE, EXCELLENT PRINTS WITHOUT A WALLET-BREAKING PRICE. The Rostock Max v2 scored well in our testing, and includes a lot of features for those willing to take the time to assemble their own printer. Just don't plan on keeping it in a small space — at 4' tall, this machine is a beast.

THE GREATEST SHOW ON RAILS

If you have never watched a delta printer in operation, you are missing out on one of the best shows in 3D printing. A delta operates by moving three identical carriages up and down along a linear rail. Each of these carriages is attached to arms, moving the center effector carrying the hot end. The Rostock Max v2's injection-molded carriages lock on solid and don't require adjustment, providing consistent printing.

CONTROL AND OPERATION

The built-in controls are up front and easily accessible, including the power switch and the SD card, which is conveniently located on the side of the LCD. The menus are easily navigated with a scroll knob that doubles as a push button.

The Bowden-style extruder removes weight from the effector, allowing the Rostock Max v2 to move around its giant build platform quickly. The inclusion of a filament loading and unloading script would be useful for dealing with the long Bowden tube, but this can easily be added with a few lines of custom G-code.

SCORING BIG

The Rostock Max v2 performed well on all of our tests but shined in the negative space tolerance test — it was one of the only machines ever to score a 5. Our large, clean, overnight print didn't even come close to maxing out the print area, despite the 8 hours it spent working.

CONCLUSION

While building and calibrating a 3D printer may be a little intimidating for those just getting started, the Rostock Max v2 is a great machine for veterans looking for large, tall build areas. ◉

> If you have never watched a delta printer in operation, you are missing out on one of the best shows in 3D printing

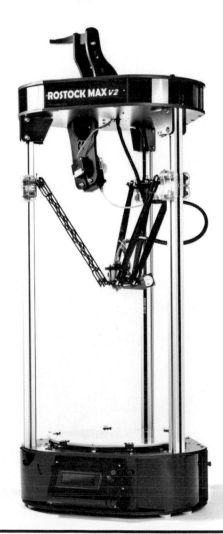

UP BOX

THIS PROFESSIONAL-LOOKING MACHINE OFFERS AN EASY OUT-OF-THE-BOX EXPERIENCE WRITTEN BY JIM RODDA AND CHRIS YOHE

THIS SLEEK MACHINE SEEMS TO TARGET THE PROSUMER MARKET WITH ITS FUNCTIONALITY AND PRICE. The Up Box is easy to use, comes with lots of extras, and includes free lifetime tech support.

SIMPLE SETUP
The proprietary printing software is a quick and easy install. Before the first print, users must manually level the print surface by alternatively turning thumbscrews and typing Z-axis values into the print software. Fortunately, this tedious process only needs to be done once: Thereafter, the printer's 9-point auto-leveling process adjusts the raft thickness to compensate for minor unevenness.

This machine includes a HEPA filter that all but eliminates the typical ABS fumes. It's usually very quiet, too — several times during our tests a visual inspection was required to confirm that the machine was still printing.

SOME HICCUPS
Print adhesion is perhaps too good, making the removal of finished prints off the perf board with a putty knife a fairly vigorous physical experience. Test prints using default settings were of average quality, and would have required significant trimming and sanding to completely remove the raft from the bottom. Great marks on the horizontal finish and

THE AFINIA H800

The H800 is the same as the Up Box, with a different color scheme and slight software differences. It includes free lifetime Afinia tech support. Afinia has scored well in the past with their other Up-licensed machines.

negative space tolerance probes gave way to less than average results on a fair number of tests, leading to a middle of the pack finish.

The bundled software feels cumbersome and clunky, with little attention paid to establishing a smooth workflow. It crashed several times while choosing resolution settings, but a call to tech support resulted in a functional work-around.

CONCLUSION
The Up Box's average print quality is a drawback for high-end design users and printing enthusiasts. The higher price point may also turn off some buyers, but those looking for an easy to use machine with a look that fits in an office or workspace might be willing to pay the premium. ✪

> Removing finished prints from the perf board with a putty knife is a fairly vigorous physical experience

PRINT SCORES 0 1 2 3 4 5

Vertical Surface Finish
Horizontal Finish
Dimensional Accuracy
Overhangs
Bridging
Negative Space
Retraction
Support Material
Z Wobble

TOTAL 27

MANUFACTURER 3D Printing Systems
PRICE AS TESTED $1,899
BUILD VOLUME 255×255×205mm
BED STYLE Heated perf board
FILAMENT SIZE 1.75mm
OPEN FILAMENT? No
TEMPERATURE CONTROL? Yes, tool head (260°C max)
PRINT UNTETHERED? Yes (Unplug USB after starting)
ONBOARD CONTROLS? Yes
HOST/SLICER SOFTWARE Up bundled slicer/print software
OS Mac, Windows
FIRMWARE Closed
OPEN SOFTWARE? No
OPEN HARDWARE? No
MAXIMUM DECIBELS 67.1

PRO TIPS
Use the buttons on the side of the printer as shortcuts to common printing tasks like initialization and bed preheating.

Be sure to keep the print bed warm between prints, as it cools quickly and may take a few minutes to warm up.

WHY TO BUY
The machine promises an easy out-of-the-box printing experience in an attractive, professional-looking body. Lifetime tech support is a huge plus, and the printer ships with plenty of extras to help users get going.

3dprintingsystems.com

Gunther Kirsch

RESULTS

PRINT SCORES 0 1 2 3 4 5

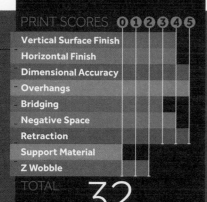

Vertical Surface Finish
Horizontal Finish
Dimensional Accuracy
Overhangs
Bridging
Negative Space
Retraction
Support Material
Z Wobble

TOTAL 32

MANUFACTURER Fusion3 Design
PRICE AS TESTED $3,975 (single extruder) and $4,975 (dual)
BUILD VOLUME 306×306×306mm
BED STYLE Heated, mirrored glass
FILAMENT SIZE 1.75mm
OPEN FILAMENT? Yes
TEMPERATURE CONTROL? Yes, tool head (300° max)
PRINT UNTETHERED? Yes (SD card)
ONBOARD CONTROLS? Yes. LCD screen with scroll wheel
HOST/SLICER SOFTWARE Simplify 3D
OS Linux, Mac, Windows
FIRMWARE Marlin
OPEN SOFTWARE? No
OPEN HARDWARE No
MAXIMUM DECIBELS 67.3

fusion3design.com

PRO TIPS

Take great care to level the build plate and experiment with build surfaces.

Fusion3 recommends using a glue stick to help prints stick, but between builds that were sticking to it and concerns over scratching the mirror, we found that the old blue tape method was more reliable.

Go easy on the scroll wheel; the model we tested was very sensitive.

WHY TO BUY

A good choice for anyone looking to make big prints quickly without compromising quality or reliability.

Gunther Kirsch

RESULTS

FUSION3 F306

THIS BIG, FUN PRINTER HAS A PRICE TAG TO MATCH

WRITTEN BY TOM BURTONWOOD

THE F306 IS A THING OF BEAUTY AND WILL DELIGHT 3D PRINTER ENTHUSIASTS. IT'S AN ENTRANCING MACHINE, with an open frame design, brightly lit build area, mirrored platform, and nylon thread movements. The printing seemed effortless and graceful, and it scored highly during our testing.

While it is not an open source machine, this Core XY bot gives Adrian Bowyer's Darwin RepRap a thorough reimagining. The tool head moves on the X and Y axes while the build platform rises up to meet it and descends as the model is built.

COOL TOOL

The excellent construction of this machine is evident in its exquisite attention to detail. The F306 benefits significantly from the E3D actively cooled all-metal hot end that comes standard. The extruded filament is cooled by another fan mounted above. Its mirrored glass bed is an interesting choice, given how easy it can be to scratch.

The overnight print (see bottom left) was a favorite with testers — the 270mm-tall job took around 11 hours to complete.

BIG MONEY

I must admit some slight sticker shock when I learned the retail price. The build volume is not that much bigger than the LulzBot Taz5, which retails for $1,700 less. Nor was I a big fan of the leveling process. Auto bed leveling is becoming more and more popular and for the price of the F306, I think it's something Fusion3 should consider for future iterations.

CONCLUSION

The machine performed very well on the test models. It ran quickly for its size, and the recommended cinnamon red filament from Atomic Filament printed nicely and looked great under the LEDs. As an artist working more and more on large prints I would certainly consider adding the F306 to my bot farm. I think this is a good choice for anyone looking to make big prints quickly without compromising quality or reliability. ◐

A joy to watch in action and a lot of fun to use

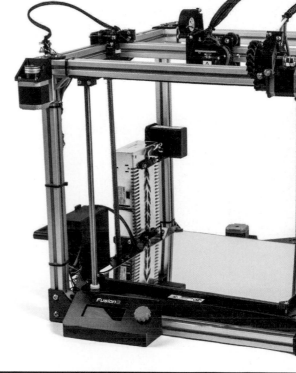

POWERSPEC ULTRA

THIS MAKERBOT HOMAGE OFFERS AFFORDABLE RELIABILITY

WRITTEN BY SPENCER ZAWASKY

THIS MACHINE PROMISES CAPABILITIES COMPARABLE TO THE MAKERBOT REPLICATOR 2 OR 2X. Similar in appearance, the PowerSpec Ultra 3D has dual extruders, an enclosed print area (including acrylic windows and removable hood), and a graphical touchscreen interface. The enclosed area and heated build platform make large, warp-free ABS prints possible. As a mid-range option it offers a balance of capability and price with decent print quality.

We were granted access to a preproduction unit, and there were some minor issues that will likely be addressed before its official release — namely, buggy firmware. But the issues were never a real impediment to printing, nor did these flaws translate to problems in the prints. Once we were up, the Ultra ran consistently and without any significant surprises. The convenient online chat support was able to get us unsnagged and going.

POWER USER

The Ultra shines in one area particularly thanks to the all-new PowerPrint software. While visually similar to MakerWare, PowerPrint eschews total web assimilation for simple usability improvements, like automatically displaying a graphical preview of the generated G-code. The software regularly produced accurate print time estimates, though the application did have its idiosyncrasies.

It was obvious that PowerPrint was relying on Slic3r for the heavy lifting — suggesting that with a little poking around, a knowledgeable user should be able to use Slic3r directly, taking advantage of all the advanced configuration options and regular software updates that this implies, and without fussing with third-party firmware.

CONCLUSION

Priced as it is, the PowerSpec Ultra 3D falls safely in the midrange of our menagerie. With clean, straightforward software and a few premium features — like a metal frame, an acrylic hood, and Wi-Fi connectivity — it punches higher than its midrange cousins, giving it an edge in value. Plus, the Ultra holds the promise of using a wide range of materials and expert software to unlock its full potential. ●

> It punches higher than its midrange cousins, giving it an edge in value

PRINT SCORES 0 1 2 3 4 5

- Vertical Surface Finish
- Horizontal Finish
- Dimensional Accuracy
- Overhangs
- Bridging
- Negative Space
- Retraction
- Support Material
- Z Wobble

TOTAL 30

MANUFACTURER Micro Center
PRICE AS TESTED $799
BUILD VOLUME 229×150×150mm
BED STYLE Heated plastic
FILAMENT SIZE 1.75mm
OPEN FILAMENT? Yes
TEMPERATURE CONTROL? Yes, tool head (300°C max)
PRINT UNTETHERED? Yes (USB and SD card options)
ONBOARD CONTROLS? Yes, small touch screen with graphical menu
HOST/ SLICER SOFTWARE Vendor provided "PowerPrint.app" (host); Slic3r — accessed through the PowerPrint software UI (slicer)
OS Linux, Mac, Windows
FIRMWARE Closed
OPEN SOFTWARE? No, but outputs G-code
OPEN HARDWARE? No, but accepts standard G-code files
MAXIMUM DECIBELS 68.7

PRO TIPS

The critical PowerPrint Software is on the included SD card. It might seem obvious, but more than one person missed it.

The blower for cooling the prints is mounted on the left, so favor the left extruder, even though the software defaults to the right extruder.

WHY TO BUY

This machine has reliable operation, decent quality, and simple software at a reasonable price. All of that comes out of the box; for the adventurous, it's a jumping off point for advanced software and materials.

RESULTS

microcenter.com

Gunther Kirsch

PRINT SCORES 0 1 2 3 4 5

- Vertical Surface Finish
- Horizontal Finish
- Dimensional Accuracy
- Overhangs
- Bridging
- Negative Space
- Retraction
- Support Material
- Z Wobble

TOTAL **13**

MANUFACTURER Polar 3D
PRICE AS TESTED $799 ($599 for students and schools)
BUILD VOLUME 203mm diameter×152mm
BED STYLE Unheated mirrored glass
FILAMENT SIZE 1.75mm
OPEN FILAMENT? Yes
TEMPERATURE CONTROL? No
PRINT UNTETHERED? Yes, Polar Cloud online service
ONBOARD CONTROLS? No
HOST/SLICER SOFTWARE Polar Cloud, with Cura settings
OS Linux, Mac, Windows
FIRMWARE Proprietary
OPEN SOFTWARE? No
OPEN HARDWARE? No
MAXIMUM DECIBELS 77

polar3d.com

Gunther Kirsch

POLAR 3D WRITTEN BY SAMUEL N. BERNIER

THE COMPANY'S DEBUT PRINTER USES A ROTATING PLATFORM THAT MOVES BACK AND FORTH and an extruder that moves exclusively on the Z-axis (up and down).

Where we were expecting more speed and precision from the polar coordinate system, instead we got extremely slow and poor prints. The quick rotations of the platform caused XY offsets in the longer prints, while the constant speed change from the pivot points also gravely affects the model's surface. We only used Polar 3D's presets, so there is hope for improvement.

SURPRISING HARDWARE

A wide-angle camera behind the nozzle gives a great print-monitoring experience, but the choice of a mirrored build plate is odd since they often get glued, taped, and scratched during use.

The clever Polar Cloud is intended to serve as a web community to share technical support, design challenges, 3D model libraries, and group projects.

CONCLUSION

It's a portable solution that won't be too heavy on the wallet while still looking great on your desk. ◐

THIS MACHINE PRODUCES POOR PRINTS SLOWLY, BUT HAS POTENTIAL

PRINT SCORES 0 1 2 3 4 5

- Vertical Surface Finish
- Horizontal Finish
- Dimensional Accuracy
- Overhangs
- Bridging
- Negative Space
- Retraction
- Support Material
- Z Wobble

TOTAL **19**

MANUFACTURER Fusion Tech
PRICE AS TESTED $1,200
BUILD VOLUME 305×205×175mm
BED STYLE Unheated acrylic
FILAMENT SIZE 3mm
OPEN FILAMENT? Yes
TEMPERATURE CONTROL? Yes, tool head (230°C max)
PRINT UNTETHERED? Yes (SD card)
ONBOARD CONTROLS? Yes (LCD screen and knob for navigation)
HOST/SLICER SOFTWARE ideaMaker
OS Mac, Windows
FIRMWARE Unknown
OPEN SOFTWARE? No, but accepts G-code
OPEN HARDWARE? No
MAXIMUM DECIBELS 63.4

ideaprinter-usa.com

IDEAPRINTER F100 WRITTEN BY MADELENE STANLEY

THIS RELATIVELY INEXPENSIVE, LARGE-FORMAT 3D PRINTER IS EASY TO USE. It's open enough for tinkering, either with the native software options, or by using another open program to create the G-code.

MODERATE TESTS, SUBLIME SOFTWARE

The F100 was able to print all of our test models with adequate success, however each had extra threads and artifacts attached to the outer shell.

All settings were prefilled in the ideaMaker software, and it had more features than anticipated. The default template settings of High Quality, Standard, and Speed all had the same 0.1mm layer height, with different infill speeds and densities.

The bed leveling procedure is intuitive, and quickly accessed through the onboard menu, which also includes a filament-loading step.

CONCLUSION

While its score doesn't reflect this, we like the F100. The combination of high resolution, fast print speeds, large build volume, and the production of great prints outside the shootout make it a machine we want to see more out of. ◐

THE F100 OFFERS A HUGE BUILD VOLUME AT A REASONABLE PRICE

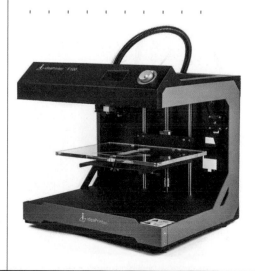

BEEINSCHOOL
WRITTEN BY CHANDI CAMPBELL

GREAT BRANDING, GOOD PRINTS, BUT MAY MISS THE MARK FOR CLASSROOMS

THIS MACHINE PRODUCES HIGH QUALITY PRINTS, CONSISTENTLY. Print slicing and machine calibration through BeeSoft are simple, reliable, and very easy to use. Marketed to students and teachers, the BeeInSchool is a slightly cheaper variation on BeeTheFirst, which tested well in last year's Shootout. The compact design, assisted calibration, and integrated handle make this a portable plug-and-play machine.

UPGRADED EXTRUDER, FINICKY FILAMENT
While the new extruder is designed to print a wider variety of PLAs, only proprietary PLA was used during testing. It ran smoothly, but a lot of pressure was required to load it without buckling and breaking, and we had a hard time unloading it even with a hot extruder and the software maintenance assist for both operations. Additionally, some steel components appeared to be collecting some rust.

CONCLUSION
The prints scored high during testing, and it seems sturdy aside from the filament issues. But we're not clear on what makes this an educational printer. ◓

PRINT SCORES 0 1 2 3 4 5

Vertical Surface Finish	
Horizontal Finish	
Dimensional Accuracy	
Overhangs	
Bridging	
Negative Space	
Retraction	
Support Material	
Z Wobble	

TOTAL 28

MANUFACTURER BeeVeryCreative
PRICE AS TESTED $1,647
BUILD VOLUME 190×135×125mm
BED STYLE Unheated acrylic
FILAMENT SIZE 1.75mm
OPEN FILAMENT? Yes, but custom spools fit the machine best
TEMPERATURE CONTROL? No
PRINT UNTETHERED? Tethered for start-up, disconnect during print (optional BeeConnect allows untethered)
ONBOARD CONTROLS? No
HOST/SLICER SOFTWARE BeeSoft (optional with BeeConnect device: Cura, Slicer)
OS Linux, Mac, Windows
FIRMWARE Marlin-based
OPEN SOFTWARE? Yes, GNU
OPEN HARDWARE? No, but a few open parts
MAXIMUM DECIBELS 63.4

beeverycreative.com

M3D MICRO
WRITTEN BY MATT STULTZ AND CLAUDIA NG

THIS TINY MACHINE COMBINES AN INTERESTING DESIGN WITH A SLOW, SLOW OUTPUT

THE M3D MICRO IS QUITE THE LOOKER — its smooth injection-molded body has almost no visible mechanics. Unfortunately, its micro print size, mediocre (at best) print quality, and shockingly slow print times don't match its looks.

PEDESTRIAN PACE
This machine is running, on average, 4 to 5 times slower than other printers — so slow, in fact, that we didn't complete the testing during our weekend. This might be forgiven if the prints were flawless, but sadly, they are not. We're hopeful, however, that the mechanics can hold up to faster speeds and the issues can be sorted out in software.

M3D has chosen to write its own slicing and host software package. It's a good start but could continue to see improvements.

CONCLUSION
We own and have access to a lot of printers, but at $350, we could see buying the M3D as a fun novelty. This isn't a good investment as a first or only printer, but if you are looking for an extra toy, the Micro could be a good fit. ◓

PRINT SCORES 0 1 2 3 4 5

Vertical Surface Finish	
Horizontal Finish	
Dimensional Accuracy	
Overhangs	
Bridging	
Negative Space	
Retraction	
Support Material	
Z Wobble	

TOTAL 18

MANUFACTURER M3D
PRICE AS TESTED $349
BUILD VOLUME 109x113x116mm
BED STYLE Unheated BuildTak
FILAMENT SIZE 1.75mm
OPEN FILAMENT? Yes, but custom spools fit the machine best
TEMPERATURE CONTROL? Yes, tool head (240 °C max)
PRINT UNTETHERED? No
ONBOARD CONTROLS? No
HOST/SLICER SOFTWARE M3D Software, proprietary
OS Windows, Mac
FIRMWARE Proprietary
OPEN SOFTWARE? No
OPEN HARDWARE? No
MAXIMUM DECIBELS 66.1

printm3d.com

Günther Kirsch

FABULOUS FILAMENTS

WHY USE REGULAR PLA OR ABS WHEN SO MANY EXOTIC ALTERNATIVES ABOUND?

WRITTEN BY SPENCER ZAWASKY

3D PRINTING WITH STANDARD PLASTICS IS AMAZING IN ITS OWN RIGHT, but printing with alternative filaments is a step up to a new level. Many of these infuse common PLA or ABS with additives, and print with normal settings. Others require much higher temperatures and slower speeds. Some may even wear out your nozzle. Still, the right material can take your next project from neat to remarkable.

Here are our favorite *other* materials this year, and a little bit about what makes them great. ✪

COLORFABB BRONZEFILL
Essentially PLA blended with powdered bronze, bronzeFill comes out looking more metal than plastic. It has noticeable added heft; it can even be polished, tarnished, or weathered to look like pure bronze. BronzeFill makes what might look like a plastic toy seem like a venerable bronzed bust.

PROTO-PASTA MAGNETIC IRON PLA
Like bronzeFill, this PLA infused with powdered iron produces a metallic result, but the grainy gunmetal finish has its own somber flair. Instead of tarnish, a real rusted look is possible. Plus, you can get this PLA to stick to magnets — a mechanical property that might be useful to inventive Makers. Be careful though, it's quite abrasive and can accelerate the wear of your nozzle.

PROTO-PASTA CONDUCTIVE PLA
Adding carbon black to PLA creates a conductive plastic that is slightly more flexible than normal. Although it has reduced layer adhesion, that's a small price to pay for charge mobility! While not quite conductive enough to be a media for making functional circuit boards, this material sticks well to PLA and is suitable for making dual extrusion structures with electrical properties.

VARIOUS WOOD FILAMENTS
There are an increasing number of wood-fiber based PLAs springing up. They offer a very convincing wood finish of various hues. However, already saturated with plastic they are not particularly thirsty for stains, varnishes, or other sink-in wood treatments. The enhanced pulchritude comes at the cost of reduced flexibility and tensile strength.

NINJAFLEX
Made from thermoplastic elastomers, NinjaFlex filament creates prints that are super stretchable, surviving punishment no ABS or PLA could tolerate. Be warned, though, this stuff has a devilish tendency to squeeze out of your extruder in directions other than though the nozzle.

TAULMAN BRIDGE
Nylons can be vexing, not sticking to anything else, causing stubborn rubbery extruder jams, and requiring very high extruder temperatures. But Taulman concocted Bridge to cross the printable-nylon gap from challenging to eminently printable. It melts at lower temperatures, extrudes more stiffly, and with a bit of glue stick on your platform, lets you print a space-age material. Super tough, mildly flexible, and impervious to glue, nylon parts can be a superior addition to any mechanical design. Bridge filament is highly susceptible to moisture, however, so keep it dry.

THE RISE OF
RESIN

WRITTEN BY MATT STULTZ

THE HOTTEST TOOL FOR MAKERS IS POWERFUL, VERSATILE, AND SPENDY — HERE'S WHAT YOU NEED TO KNOW TO GET STARTED

Uni-rex – thingiverse.com/thing:325809
printed at Formlabs by Colin Raney

FUSED FILAMENT MACHINES MAY CONTINUE TO DOMINATE THE 3D PRINTING MARKET, BUT THOSE LOOKING FOR the highest-quality prints turn to resin. Although growth of this segment hasn't been as fast, we tested more resin printers this year than any other — a sign that their potential is finally maturing.

There are two major types of resin-based printers: the traditional stereolithography (SLA) printers, which use a highly accurate laser beam to trace a path along light-sensitive resin to create each layer; and digital light processing (DLP) printers, which, rather than tracing, use off-the-shelf home-theatre projectors to cast and cure a full slice of your model all at once. Both of these machines are often lumped under the same SLA title.

Laser-based printers still dominate resin printer offerings. The high precision and fast, reliable cure times from the laser make for great machines. The only drawback is that the mirrors and drivers (known as galvanometers) can be costly and difficult to manufacture, leading to a generally higher price.

DLP endeavors have opened the door for low cost, DIY resin printers that can compete with industrial machines for print quality. Since the projector is capable of shining an image over the full build area, the only mechanics needed for a small printer are a single Z-stage motor and slide.

One of the biggest advancements this year is the new resin formulations coming to market. Where fairly brittle resins were previously the only option, you can now find resins for lost-wax style casting, flexible applications, and extra-strength needs.

For our SLA reviews, we used a variety of test models to determine print quality; the Make: Rook continues to be our go-to model.

With a few exciting new machines on the horizon, and at least some of the legal issues that have plagued this class of printers in the past clearing up, this could be a great year to get started with an SLA of your own.

SLA PRINTER PROJECTS:

Printing Optical Lenses

One of the most intriguing uses of transparent SLA is in optics. Formlabs shows that printing your own eyeglasses is within reach. makezine. com/go/3d-printing-optical-lenses-with-formlabs

DIY Wedding Rings

SLA is the top choice for prototyping jewelry. Here's how to design your own rings, then get them cast in precious metals. makezine. com/go/direct-digital-

o/fab-factory

…ERATION

…STULTZ

Matt Stultz

OS
Windows, Mac

FIRMWARE
Proprietary

OPEN SOFTWARE?
No

OPEN HARDWARE?
No

PRO TIPS
When ordering alternative resin formulations, be sure to order extra vats too. This will prevent cross contamination and make swapping faster.

WHY TO BUY
The upgraded build volume, self-filling resin vat, and Wi-Fi printing keeps Formlabs on the top of the SLA pack.

RESULTS

…SER, SLEEKER, AND STREAMLINED

The most obvious change to the Form 2 is the increased build volume, but the mechanics have also been upgraded: A horizontal slide peel, combined with a vat wiper to remove debris, has replaced the original hinged peel for separating new layers from the bottom of the resin vat. The printer also possesses a more powerful laser and new custom galvanometers for faster cure times and greater precision, and protective glass keeps components dust- and resin-free.

REVAMPED RESIN BOTTLE
An integrated, chipped bottle recognizes what type of resin is being installed and keeps track of how much is left, then automatically fills the vat with a system that leaves no moving parts or tubes in contact with the resin. Of course, because chipped proprietary materials can throw up red flags, Formlabs also includes an open mode, which doesn't auto fill, but allows a user to pour in any resin they want.

…e what we …prints that run …rints, but still …om the laser …with was an … is striving to …er when they

…e SLA game, …ct to get. You can get started at a lower cost, but the Form 2's features are currently unbeatable and are especially powerful for collaborative environments. ✐

Formlabs has taken all they have learned and released a machine that is set to keep them on top

Print Complete

formlabs

NOBEL 1.0

THIS MACHINE IS A PROMISING OPTION FOR HIGH-QUALITY PRINTS WITHOUT THE HIGH PRICE WRITTEN BY MATT STULTZ

IT IS IMPOSSIBLE TO REVIEW THE NOBEL 1.0 WITHOUT COMPARING IT TO THE FORMLABS FORM 1+. They have similar styling and mechanics, similar build volumes, and use similar resins.

One place the Nobel distinguishes itself is its $1,499 price tag. Having spent time with both printers, it was easy to see where XYZ cut down on cost: The Nobel abandons the Form 1+'s elegant curves for seams and joints, and the Form's cleverly hinged acrylic lid for an unattached one that is removed manually.

AUTOMATED RESIN FILLING

In fairness, the Nobel also has a few features its pretty competitor is missing. The machine includes onboard controls, a USB thumb stick to slice and transfer models without a computer, and uses a series of tubes to automatically (although noisily) fill the vat from an open bottle of resin.

To remove the build plate, the lid must be fully removed vertically — making it tough to use on a cramped workbench or under a shelf — and the USB port for the memory stick is on the machine's back where it is hard to access.

Like all other XYZ machines, the Nobel 1.0 won't allow you to use just any raw materials — the resin is chipped with an NFC tag. The only upside: XYZ's resin is still cheaper than those sold by many of its competitors.

SHAKY SOFTWARE

The Windows-only software is still under development and feels that way. Features are changing and many don't feel fully baked.

In terms of print quality, I found that finely detailed objects printed well, but larger and more solid objects were not always fully cured. Some had liquid pockets just below the surface, which would break open, leaving a gooey, ruined print. This feels like a software issue, and hopefully it will be worked out as they move forward with development.

CONCLUSION

The Nobel 1.0 addresses many issues with the Form 1 and 1+ but now that the Form 2 has come to market filling in these holes and more, the Nobel 1.0 might have missed its chance to be a leader. With its great feature set and comparatively low price tag, the Nobel 1.0 is still a formidable competitor so long as XYZ continues to improve the software. ◉

> Like all other XYZ machines, the Nobel 1.0 won't allow you to use just any raw materials — the resin is chipped with an NFC tag

MACHINE RATING

0 1 2 3 4 5

- EASE OF USE
- SOFTWARE SIMPLICITY
- CONSISTENCY
- RESIN OPTIONS
- COVET FACTOR

MANUFACTURER
XYZprinting

PRICE AS TESTED
$1,499

BUILD VOLUME
128×128×200mm

PRINT UNTETHERED?
Yes (USB stick)

OPEN RESIN:
No, requires NFC chipped resin

ONBOARD CONTROLS?
Yes, an LCD with navigation buttons

HOST/SLICER SOFTWARE
XYZWareNobel

OS
Windows only

FIRMWARE
Closed

OPEN SOFTWARE?
No

OPEN HARDWARE?
No

xyzprinting.com

PRO TIPS

Buy some extra gloves and safety glasses, as the ones included will go quickly.

Assemble a stand to hold the bed in place when removing prints, and rig up a second alcohol bath for soaking prints.

WHY TO BUY

If you are looking to get high-quality SLA prints without breaking the bank, the Nobel 1.0 is a promising option. Further software improvements could make this a big winner.

Kelly Egan

RESULTS

MACHINE RATING

0 1 2 3 4 5

- EASE OF USE
- SOFTWARE SIMPLICITY
- CONSISTENCY
- RESIN OPTIONS
- COVET FACTOR

MANUFACTURER
Kudo3D

PRICE AS TESTED
$3,208

BUILD VOLUME
192×108×243mm

PRINT UNTETHERED?
No

OPEN RESIN:
Yes, compatible with third-party resins

ONBOARD CONTROLS?
No

HOST/SLICER SOFTWARE
Creation Workshop

OS
Windows only

FIRMWARE
Yes, Marlin

OPEN SOFTWARE?
Yes

OPEN HARDWARE?
No

kudo3d.com

PRO TIPS

Remember to prep for the target XY resolution and build volume for your part before adding resin into the PSP vat. Kudo3D community member Jensa contributed a handy printable sizing chart to speed up this process.

WHY TO BUY

By leveraging the flexible PSP resin reservoir on a grand scale, Kudo3D eliminated several costly moving parts, making this modular, hackable, DIY machine a good value for seasoned users looking to experiment with big builds and new resins.

RESULTS

Matt Stultz

TITAN 1

GOOD VALUE FOR A MODULAR, HACKABLE MACHINE, BUT A CHALLENGE FOR BEGINNERS WRITTEN BY MATT GRIFFIN

THE TITAN 1 IS A SERIOUS TINKERER MACHINE. It's perfect for those who already have a preliminary background in SLA and look forward to the prospect of leveraging passion, puzzle-solving prowess, and community-curated knowledge to gain access to a massive, capable SLA DLP printer for thousands of dollars less than what you'd expect to pay for the build volume.

UNCONVENTIONAL CONTAINER
The included PSP resin container departs from the typical hard-sided type by introducing a softer silicone/PDMS base with two teflon side panels that flex (without spilling resin) during the "peel" process, when the Z lift pulls the latest cured layer away from the bottom of the vat. This results in shorter cycle times when the build stage dips back into position for the next layer, but is not without its downside — be prepared to spend more time calibrating and troubleshooting platform adhesion-related issues. Don't expect big prints right away.

EASY ASSEMBLE AND EASY ACCESS
The Kudo3D team's step-by-step videos make assembly of this machine a snap. The THK industrial-grade linear Z-stage mounts to the top back of the case, and drives the build plate up and down into the resin reservoir below. An easy-to-assemble acrylic shell covers the Z-stage tower and PSP container to shield the photosensitive printing compounds from exposure to ambient light. However, I suggest that those who plan to dismantle the shell for shipping or local transport should come up with their own clamps and hardware instead of using the acrylic-binding tape.

The removable, brushed-aluminum side panels can be quickly pulled off and snapped back into place, giving the operator full access to the electronics and projector inside.

FORUMS ARE YOUR FRIENDS
While setting up the machine was speedy, Kudo3D's incomplete documentation left me frustrated. Expect to supplement official materials with extensive forum hunting. The responsive community helps to round out the resources not yet officially provided.

CONCLUSION
This might be the machine for you if you are primarily interested in an experimenter's platform for learning more about DIY resin printers' function and the many Maker-friendly, third-party resins. ◉

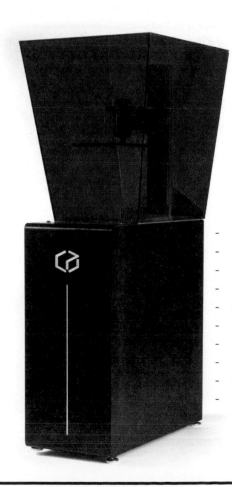

> The Kudo3D team's step-by-step videos make assembly of this machine a snap

LITTLERP

THIS INEXPENSIVE KIT IS A GREAT INTRODUCTION FOR THE SLA-CURIOUS WRITTEN BY CHRIS YOHE

THE LITTLERP COMES AS A BYOP (BRING-YOUR-OWN-PROJECTOR) RESIN PRINTER kit and I'm happy to report that it is the easiest I have assembled. The instructions (currently a series of photo galleries) leave a little to be desired for novice users, but they were sufficient and it took only an afternoon to build.

After building the main stand, you are taken through the process of setting up and calibrating a video projector. Recommended projector models can be found for between $300 and $600, so factor that into the price if you don't already have one.

SAFETY FIRST
The first few prints I ran were nothing short of amazing compared to the FDM machines I was accustomed to. Right out of the box, with only standard settings and basic configuration, the print quality lived up to the hype. Fine lettering, columns, and channels printed with ease.

That being said, safety is a serious matter. You will want to make sure to have plenty of disposable gloves, and eye protection. Also

necessary: alcohol for cleanup, and a post-print curing station (some people build their own UV light box for this, others use sunlight).

WORTH THE EFFORT
For me, the chemical hassles are outweighed by the print quality and relative print speed. The downside is the overall volume — the build area fits within a standard petri dish, so don't expect to print large-format creations. Resin costs are also higher than those of filament, which is why hollowing your designs and conserving resin are part of the learning curve.

CONCLUSION
The LittleRP is a cost-effective way to get into making high-quality prints, especially if you already own a projector. Due to the safety concerns and the slightly more open nature of this machine, using this in a classroom or on a workplace desktop would likely be out of the question. Jewelry buffs, artists, modelers, and tabletop gamers will all have a field day with this though. ◉

> **I'm happy to report that it is the easiest printer kit I have assembled**

MACHINE RATING

0 1 2 3 4 5

EASE OF USE
SOFTWARE SIMPLICITY
CONSISTENCY
RESIN OPTIONS
COVET FACTOR

MANUFACTURER
LittleRP
PRICE AS TESTED
$599
BUILD VOLUME
60×40×100mm
PRINT UNTETHERED?
No
OPEN RESIN:
Yes, compatible with 3rd-party resins
ONBOARD CONTROLS?
No
HOST/SLICER SOFTWARE
Creation Workshop
OS
Windows/Linux
FIRMWARE
Grbl
OPEN SOFTWARE?
Free to use for end users, even commercial. No open source yet.
OPEN HARDWARE?
Grbl is open source. LittleRP Arduino shield is not. Design Files have been released with non-commercial license. Expected to be released totally open in the future.

littlerp.com

PRO TIPS
Head to buildyourownsla.com — which hosts a LittleRP Specific sub-forum and is run by the Creation Workshop folks. A lot of helpful information can be found here.

WHY TO BUY
Incredible SLA print quality at an affordable price. This is one time the race to the bottom on Kickstarter got it right.

Kelly Egan

RESULTS

One little tool lets you do
great things

Michael Fogleman of Raleigh, NC, used a Handibot® Smart Power Tool to create a wooden map of all 100 North Carolina counties. The computer-driven Handibot allowed Michael to cut out the counties, carve their names, and cut the outlined base for the map. Handibot is great for creative projects of all kinds, because it enables wood cutting, carving, and milling; it can also be used to cut MDF, plastics, foam, and aluminum with engraving-level precision and impressive power.

The Handibot cuts through its base into whatever material you're working on. Accessories such as a 3D Rotary Axis, Laser Sight, and Large Material Jig allow you to expand Handibot's capabilities.

Read about Michael's process: medium.com/@fogleman

What would you like to make?
Pick up a Handibot and make it happen!

"Cutting out the county pieces was easy. I just picked up the Handibot and placed it directly onto the wood." – Michael Fogleman

1. Cut the base with a bit of tiling   2. Cut out the parts easily   3. Carve the names precisely   4. Stain, glue, display with pride

carve *your* path

CNC AND 3DP SOFTWARE GLOSSARY

MAKING SENSE OF ALL THE FABRICATION ACRONYMS

WRITTEN BY JOHN ABELLA

WITH THE RAPID EXPANSION of hobbyist-class digital fabrication equipment and the always-plummeting price points, new enthusiasts step into a world of acronyms, software packages, and terminology to master. Here's a cheat sheet to get you started.

CAD (3DP / CNC)

Computer-aided design is software that allows the user to create models in 2D or 3D formats. While CAD was initially common only in architecture and manufacturing, enthusiast-oriented packages are now readily available at low (or no!) cost.

CAM (CNC)

Computer-aided manufacturing software generates the toolpaths (using G-code) for CNC milling and cutting machines. It takes popular 2D file formats and allows the user to determine which parts of a design need to be milled or cut, how fast, at what depth, and any other details.

G-CODE (3DP / CNC)

The language used to instruct CAM systems to perform operations. In modern usage, this is almost exclusively generated by software, and not written by hand. The specific G-codes control motion, speed, rotation, depth, and other related switches and sensors used in the operation of a machine.

G-CODE SENDER (3DP / CNC)

Once G-code is generated, this software package streams the actual commands to the machine (usually via USB) to be run. While individual G-code sender packages are still used in some of the open source toolchains, many of the commercial packages now combine a slicer and a G-code sender into a single package.

MESH EDITOR (3DP / CNC)

Once a 3D model is converted to an STL format, the resulting layout of triangles is often referred to as a "mesh." A mesh editor lets the user directly edit the points of the mesh: stretching, shrinking, smoothing, or otherwise manipulating the actual shell of the 3D model.

OPENSCAD (3DP)

A software package for programmatically generating 3D shapes, complex systems of pieces, and even parametric designs. As opposed to traditional CAD programs, there is no "drawing" in OpenSCAD — all designs are defined as text and then compiled to see the resulting shapes.

SLICER (3DP)

Additive manufacturing works in layers and a "slicer" is the software package used to cut a 3D model into flat layers that can be printed one at a time. The output of a slicer is G-code that controls the path, speed, and temperature of the printer. There are both closed and open source slicer software packages.

STL (3DP / CNC)

The most common file format for 3D printing, and one that's become more popular for "2.5D" milling, where just one face of the design is carved out due to the limitations of a 3-axis machine. The STL file format represents a 3D object by describing the surface as a series of triangles. The file type is "unitless" — 1 unit could be 1mm or 1" (or any other arbitrary measure) — so it's important to know how the file was generated. ⊘

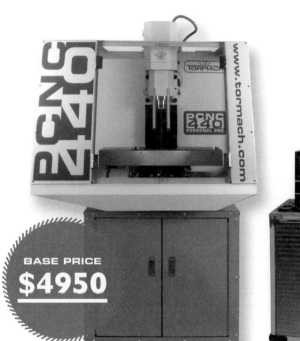

A REPRAP FAMILY TREE

TRACKING THE OPEN SOURCE DESKTOP PRINTERS THAT STARTED IT ALL

WRITTEN BY CHANDI CAMPBELL

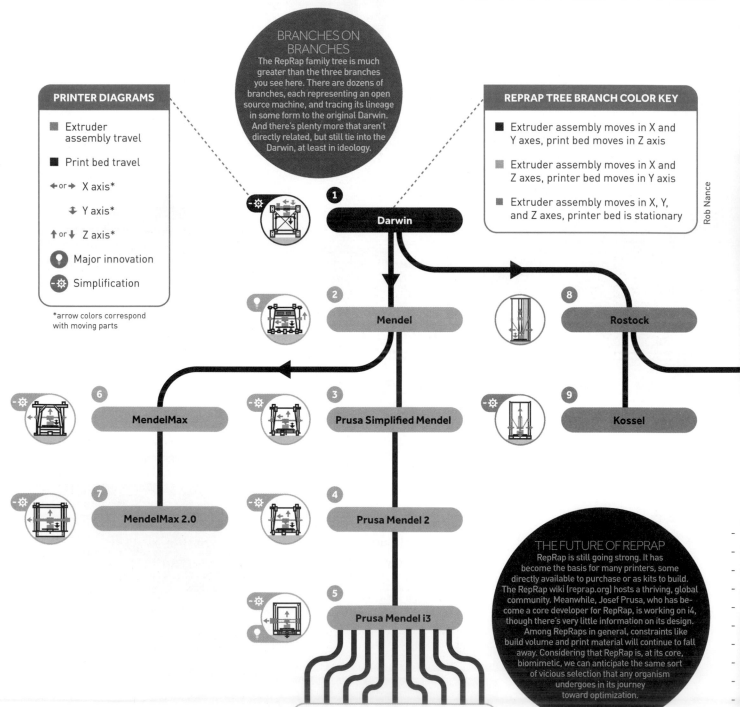

BRANCHES ON BRANCHES
The RepRap family tree is much greater than the three branches you see here. There are dozens of branches, each representing an open source machine, and tracing its lineage in some form to the original Darwin. And there's plenty more that aren't directly related, but still tie into the Darwin, at least in ideology.

PRINTER DIAGRAMS

- ■ Extruder assembly travel
- ■ Print bed travel
- ← or → X axis*
- ↕ Y axis*
- ↑ or ↓ Z axis*
- 💡 Major innovation
- ⚙ Simplification

*arrow colors correspond with moving parts

REPRAP TREE BRANCH COLOR KEY

- ■ Extruder assembly moves in X and Y axes, print bed moves in Z axis
- ■ Extruder assembly moves in X and Z axes, printer bed moves in Y axis
- ■ Extruder assembly moves in X, Y, and Z axes, printer bed is stationary

Rob Nance

1 Darwin

2 Mendel

8 Rostock

6 MendelMax

3 Prusa Simplified Mendel

9 Kossel

7 MendelMax 2.0

4 Prusa Mendel 2

5 Prusa Mendel i3

THE FUTURE OF REPRAP
RepRap is still going strong. It has become the basis for many printers, some directly available to purchase or as kits to build. The RepRap wiki (reprap.org) hosts a thriving, global community. Meanwhile, Josef Prusa, who has become a core developer for RepRap, is working on i4, though there's very little information on its design. Among RepRaps in general, constraints like build volume and print material will continue to fall away. Considering that RepRap is, at its core, biomimetic, we can anticipate the same sort of vicious selection that any organism undergoes in its journey toward optimization.

Many i3s

REPRAP IS SHORT FOR REPLICATING RAPID PROTOTYPERS, THE ORIGINAL OPEN SOURCE DESKTOP 3D PRINTERS.

Initially conceived in 2005 by Adrian Bowyer, at the time a senior mechanical engineering lecturer at the University of Bath in England, RepRaps were designed to reproduce many of their own components — printers printing printers. The rest of the hardware needed is easily acquired at a low cost, allowing everyone equal access to the necessary technology to produce their own goods.

Bowyer went on to design the first well-known RepRap, the Darwin. Now a director at RepRapPro, he describes the RepRap project as a symbiotic relationship: These machines would produce our goods at a low cost while we would provide the assistance they needed to reproduce themselves, like the relationship between insects and the plants they pollinate. RepRaps would undergo a process of selection similar to biological organisms. They would evolve as they spread and as new technology became available.

And that is exactly what has happened. From that concept grew — and continues to grow — a great family of printers bearing a great many innovations that appear not only throughout the RepRap tree but in commercial printers. Many of the printers reviewed in this issue owe parts of their design to the RepRap movement. ◑

Rostock Mini

A REPRAP FOR EVERYONE
With the i3's popularity many companies have created their own version of this capable printer that you can buy today. BQ, BeeVeryCreative, MakerFarm, MakeBlock, and Prusa3D, to name a few, all produce i3 variants as kits or fully assembled packages. With so many options, there may be an i3 to fit everyone.

MIDWEST REPRAP FEST
The biggest RepRap event in the world takes place each year in Goshen, Indiana, home of printer company SeeMeCNC. Over the weekend, attendees check out custom and rare machines, and attend talks by RepRap luminaries. It's one of the best places to see how large, diverse, and creative the RepRap community is.

OPEN SOURCE
A passionate and robust community, freely exchanging ideas, is pushing the evolution of this symbiosis forward. Cross-pollination of designs is leading to variation that is already pretty complex.

At the heart of RepRap is its open source GNU General Public License. Originally created for software, GPL is applicable here because RepRap designs and files are considered the "source code" for future iterations. Not only can you build your own, you can sell them as kits or completed machines. Just don't try to use it in a proprietary system.

Darwin line (top)

DARWIN, Adrian Bowyer, 2007 ①
The cube-shaped Darwin is considered the first RepRap printer. A Cartesian machine constructed largely of steel rods and plastic 3D printed junctions, it used a descending build platform that tended to bind due to four lead screws that raised and lowered each corner of the platform. This axial orientation was used in at least nine of this year's tested printers.

MENDEL, Ed Sells, 2009 ②
The RepRap line took a big design shift with the Mendel. In addition to being easier to assemble, the cube was traded in for a prism shape, which moved most of the weight to the base, making it more stable. Furthermore, it enabled the four lead screws to be replaced by just two, greatly reducing the binding and leveling issues that caused so many jams.

Prusa line (center)

PRUSA SIMPLIFIED MENDEL, Josef Prusa, 2010 ③
Josef Prusa, a young "proud dropout" from the University of Economics in Prague, simplified the Mendel so much that assembly time was dramatically reduced. At this stage, an experienced builder could construct one in a weekend. Fewer parts and the streamlined design meant you could print the necessary joiners in half the time.

PRUSA MENDEL 2, Josef Prusa, 2011 ④
With a few more tweaks, the second iteration of the Prusa Mendel introduced more complexity to the machine, while streamlining the hardware used to build it. A heated build plate helps prints adhere and avoid warping, and the X carriage was redesigned to accommodate linear bearings and bushings. By this point, the community was gaining traction, and this model became extremely popular, attracting many new participants and hobbyists.

PRUSA MENDEL I3, Josef Prusa, 2012 ⑤
The i3 became still more popular, a flagship for the RepRap movement. A whole tree sprouts off from here — if you're looking at getting a RepRap, it's probably an i3. Primary improvements include an open design to maximize build volume, a stable, rigid frame, and an overall reduction in parts.

MendelMax line (left)

MENDELMAX, Maker's Tool Works, 2011 ⑥
The folks at Maker's Tool Works took the Mendel concept and designed a new frame, constructing it with aluminum extrusions as opposed to rods. This made the frame very rigid and durable while retaining the original RepRap construction ideal of using printed joinery.

MENDELMAX 2.0, Maker's Tool Works, 2013 ⑦
There's a bit of crossover here; the 2.0 was clearly influenced by the Prusa i3 in its open framework, again built with rigid aluminum extrusions. This MendelMax is actually sort of a Mendel/i3 hybrid because it still has the top-mounted Z-axis motors. It features a larger print volume and a simplified construction, requiring fewer parts.

Rostock line (right)

ROSTOCK, Johann C. Rocholl, 2012 ⑧
The Rostock is a prototype delta-style RepRap and is progenitor to a far-reaching branch of the RepRap tree. Delta robots (of the non-printing variety) have a long-standing reputation for speed in various manufacturing applications, so it made perfect sense to adopt this style of bot in 3D printing. It's not unreasonable to expect print speeds of around 350mm/s from a delta printer (see SeeMeCNC's Rostock Max v2 kit, page 41).

KOSSEL, Johann C. Rocholl, 2012 ⑨
Later in 2012, Rocholl released the Kossel, another member of the RepRap family with the namesake of another famous biochemist/geneticist. (Notice the evolutionary trend?) Today, Rocholl still has the design classified as "experimental." This machine took advantage of aluminum extrusions in its frame construction, just like the MendelMax, which allowed an exceptional build height of about 400mm (16").

ROSTOCK MINI, Brian Evans, (2012) ⑩
The Rostock Mini is a shorter and more structurally stable version of the original Rostock. The other main improvement is the attractive frame components, which were designed to be CNC cut from wood or acrylic. This iteration has proven to be reliable and is very popular in the RepRap community.

SUBTRACTIVE **FABRICATION** STEPS UP

8 PORTABLE CNC MILLS AND ROUTERS, TESTED AND REVIEWED

WRITTEN BY LUIS RODRIGUEZ

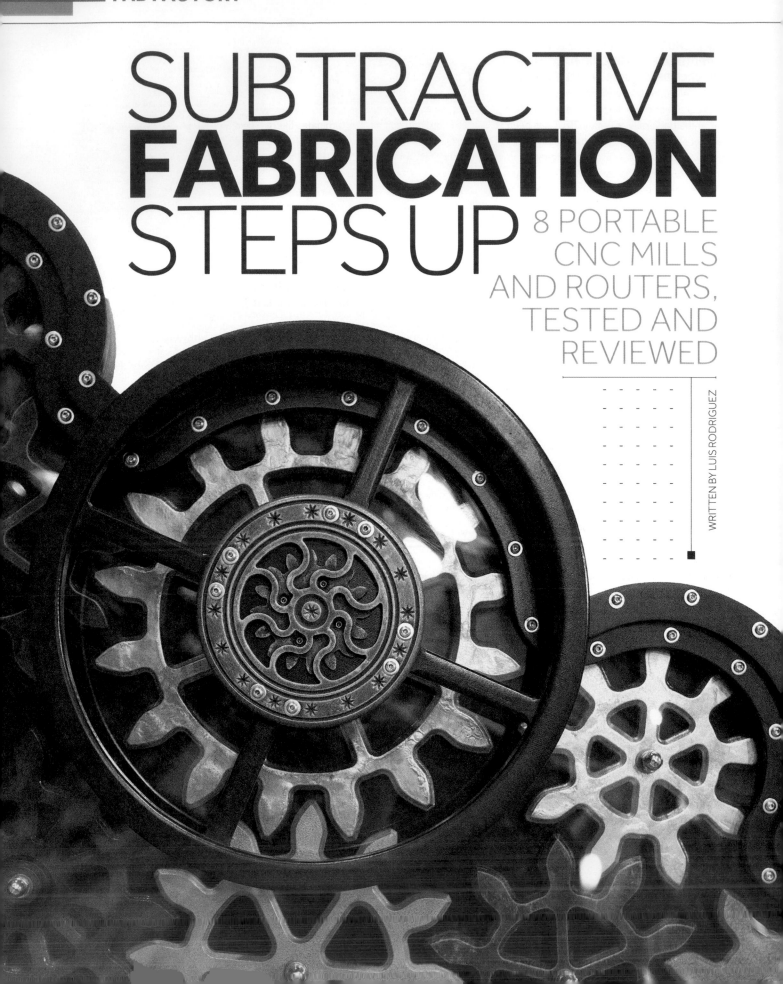

IN THE FABRICATION WORLD, CNC (COMPUTER NUMERICAL CONTROL) MACHINES HAVE BEEN AROUND FOR MORE THAN 50 YEARS, showing up over two decades before the advent of 3D printing in 1984. Over that time, the term "CNC" has been adopted as a shorthand reference to milling or routing, where a spinning end mill — a close cousin to the drill bit — chews through material to engrave, cut shapes out of flat stock, or carve intricate 3D designs from thick materials.

The transition of CNC mills and routers from heavy industrial machinery to affordable desktop tools has come at a slower pace and without the fanfare that 3D printing has experienced over the past few years. That's started to change of late, however, and there is now a variety of easy-to-use, workbench-sized cutters on the market.

MAKE ANYTHING

One of the many reasons CNC mills are so important to every makerspace, hackerspace, and fab lab is that they provide an easy way for individuals, students, and entrepreneurs to precisely produce complex physical parts for almost every project type imaginable. Woodworking, finished goods, furniture, musical instruments, signs, boats, jewelry, and more — with the power of CNC you can generate parts with accuracy and at a scale that is not possible otherwise. The CNC moniker means that the instructions that control the machine and toolpath are preprogrammed into code. It ensures that the 100th part produced is identical to the first.

The new crop of Maker-focused tools range from CNC micro mills used for cutting circuit boards to large routers that turn 4'×8' sheets of lumber into furniture. All are controlled in the very same way; the only differences are in which toolchain is used to get to the finished part.

COMPUTER MOVEMENT

Like 3D printers, which use CNC to control movement of an extruder head, the majority of CNC mills are 3-axis machines that move the toolhead in the X (left-right), Y (forward-back), and Z (up-down) directions via a belt or lead screw. Some function by moving just the spindle, others move the bed, and some move both the spindle and the bed, depending on the axis.

MATERIAL CHOICES

A CNC mill allows you to quickly and accurately work with harder materials that are not normally possible with other technologies. These can vary from hard and soft woods, laminates like plywood, plastics of any kind, and hard and soft metals. The only limitation to what you can cut is the scale and rigidity of the machine and the end mill being used.

The hardness of the material being cut and the size of the end mill makes correct setting of these instructions crucial. Too fast a setting can result in broken bits. Too slow and you risk burning your material. Set the pathway incorrectly and you could find yourself cutting right through the platform of your machine. (This is somewhat expected, though — it's called a wasteboard for a reason.)

PROGRESS THROUGH INNOVATION

With the machines and software that have come out lately, desktop CNC has gotten a whole lot easier to use than the traditional industrial CNC mills. One of the most useful recent developments involves combining the CAD/CAM functionality into one program — using one piece of software to not only design your part, but also produce the G-code. Autodesk's Fusion 360 has become a popular application for this, as well as MeshCam, MakerCam, VCarve Pro, and Aspire.

Going a step further, these programs are now being run in the cloud with nothing but a small USB utility connecting the machine to the web-based CAD/CAM solution. Inventables' Easel is already available for this, while other companies are rumored to be pursuing it as well.

Another convenient new trend is to move the computer and machine control onboard, meaning the microcontroller and computer are inside the CNC machine; with these you can wirelessly log in and connect to the machine to perform all the functions from just about anywhere.

All these innovations are designed to make it easier for the beginner and expert. Your only decisions are the size you want to work at, and how much money you are comfortable investing. Let's turn the page and figure out what's right for you. ◉

Kids Interactive Gear Wall — Design and Fabrication for the Hattiesburg Zoo in Mississippi by Because We Can.

Jillian Northrup

CNC ROUTER PROJECTS:

DIY Star Wars AT-AT

At over a foot tall, it's perfect for holding your figurines. And when it crashes into the snow, well, just carve yourself a new one. makezine.com/go/cnc-at-at

DIY Vacuum Hold-Down Table

Lose the screws and clamps — let the table hold your workpiece with suction. makezine.com/go/universal-vacuum-hold-down

CRAWLBOT

WRITTEN BY JASON LOIK

THIS REVOLUTIONARY MACHINE GIVES SUPERSIZE RESULTS AT A HOBBYIST PRICE

printrbot.com

MANUFACTURER
Printrbot

PRICE AS TESTED
$3,999

BUILD VOLUME
1,219×2,438×50mm

HOST SOFTWARE
Custom Printrbot CNC
control software

CAM
Fusion 360 or custom
Printrbot software

FIRMWARE
TinyG

OS
Cross platform
Chrome extension

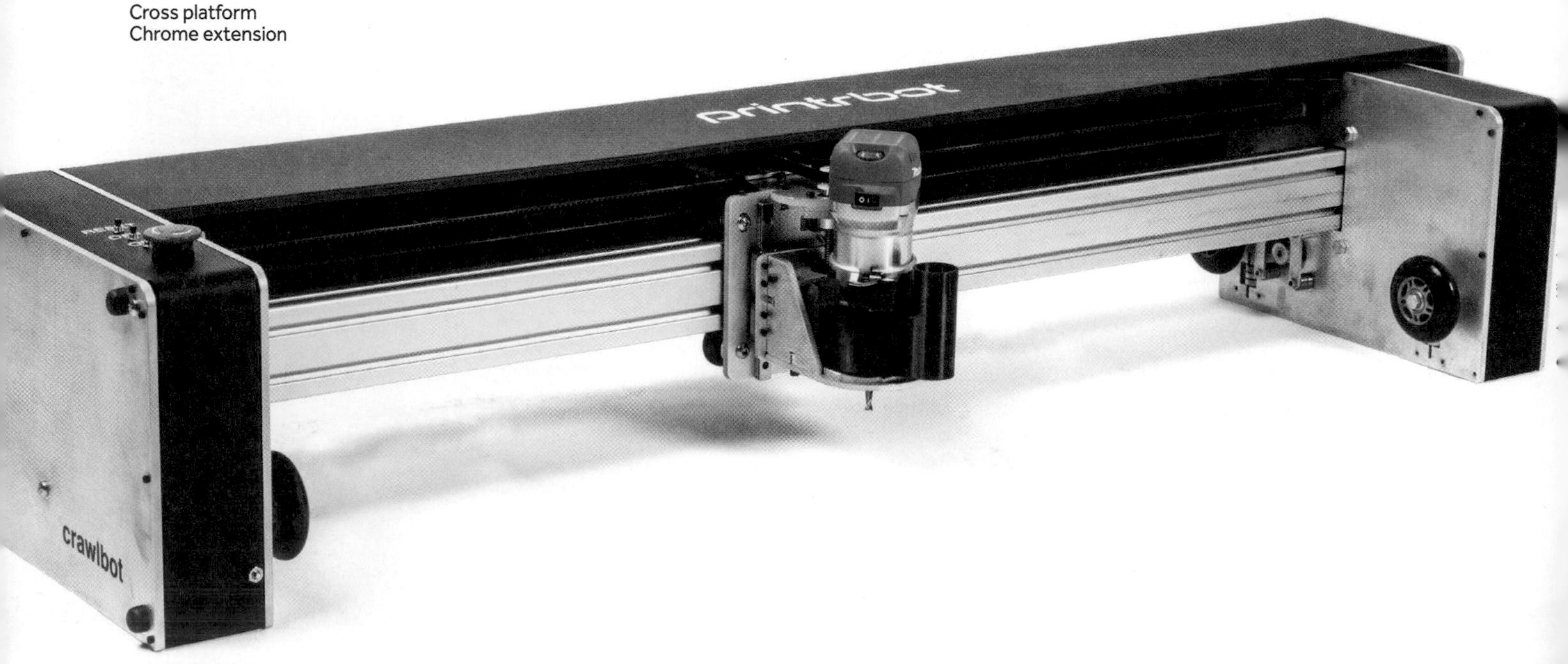

THIS HIGHLY ANTICIPATED MACHINE FROM PRINTRBOT DOES NOT DISAP- POINT. It's a game changer, marking the beginning of a new movement for CNC technology. The Crawlbot we tested was the only one in existence, the prototype. So we are happy to bring you, maybe not a full review, but a sneak peak of this revolutionary idea.

INGENIOUS ENGINEERING

The Crawlbot takes a fundamental limitation of CNC machines and throws it out the window: Traditionally, if you wanted to cut a large piece of material, then your machine would need to have an even larger frame. The Crawlbot, however, is the size of a golf bag, but can handle a full sheet of plywood.

It is ingeniously simple: Instead of relying on a normal CNC machine's oversized frame, this parasitic bot gains structure by simply attaching to the very material it's cutting. The X- and Z-axis are fairly conventional, running along sturdy extruded aluminum, but the magic is in the Y-axis. There are 2 drive belts that run the length of the wood and clamp securely to all four corners. Riding on what amounts to rollerblade wheels, and using the straight edge of the plywood as a guide, the Crawlbot lives up to the name by pulling itself along the material it is cutting.

This is the perfect machine for the at-home tinkerer, someone who can back their car out of the garage on the weekends. You can even transport it in your trunk! Printrbot states that you can simply strap it to a single piece of plywood on a few saw-horses and let it run. And in our tests, that worked perfectly, cutting out my AtFab chair with spot-on joinery.

SMOOTH AND SOLID

Keeping in mind we had the prototype, I am very impressed with the hardware. It is solidly constructed with extruded aluminum and carries a manually controlled Makita router attachment (glad to see that they didn't "cheap out" on brand — but it would be nice to have automated RPM). The cutter-head travel is elegantly smooth, driven by the fantastic TinyG motion controller housed in a sturdy box. The machine weighs 65 lbs, not unmanageable alone but it's nice to have a buddy to help for the first few runs. After that, handling setup and teardown becomes a little easier alone.

Unfortunately, the machine's greatest strength also becomes its weakness. Since the Crawlbot rides along the very surface it's cutting, it needs material to hold it as it works. This causes the bot to leave an almost 4" border, on all sides, of unusable surface. So all those wonderful designs that fit into a standard 4×8 sheet aren't compatible. I found a few workarounds for this but nothing completely satisfying. Printrbot is working on solutions to this as well.

CONCLUSION

The Crawlbot is not a complete replacement for the conventional CNC router. It's restricted to working on a fairly flat, straight-sided material. The Z-height is a bit lacking, just a little over 2 inches. Also keep in mind that the machine will set up quickly, but power cords, vacuum hoses, and data connection makes it a little less versatile than at first glance.

But because of price and portability I think this humble little machine will pull a good portion of larger CNC sales. At $4,000, it costs a fifth of what many full-sized CNC's go for. Printrbot has a great idea here. In fact it's so great (depending on what patents they have been able to secure) I think we will soon see other companies coming out of the woodwork with similar products. ◉

> ## The Crawlbot takes a fundamental limitation of CNC machines and throws it out the window

We cut and assembled this AtFab chair with the Crawlbot in just a few hours, to show the large-scale CNC capabilities this machine offers.

SAMPLE CUT

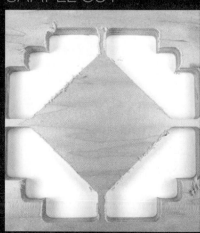

NOMAD 883

THIS STURDY, ENCLOSED MACHINE IS BOTH PRETTY AND PRECISE

WRITTEN BY SHAWN GRIMES

carbide3d.com

MANUFACTURER
Carbide 3D

PRICE AS TESTED
$2,599

BUILD VOLUME
203×203×76mm

HOST SOFTWARE
Carbide Motion

CAM
MeshCam

FIRMWARE
Grbl

OS
Mac, Windows

Available at
Maker Shed

Gunther Kirsch

FROM THE SAME COMPANY THAT CREATED THE SHAPEOKO 3, THE NOMAD 883 IS TOUTED AS AN ALTERNATIVE to their CNC machine kits. It comes as a fully assembled and enclosed machine with the option of two different looks: bamboo or HDPE.

METICULOUS 3D MILLING

The setup process for the Nomad 883 was really simple. After downloading the Carbide Motion software, I followed the first tutorial on their site and in only about 15 minutes I had a milled wrench thanks to the provided G-code file.

This early success made things exciting. The next tutorial was using MeshCam to do a 3D carving of an STL file and again it was a great success and incredibly inspiring. After the testing, it was clear that this is the Nomad's primary use case, carving STLs into materials that a 3D printer just can't handle, and with a precision that 3D printers can't match.

When you use a MeshCam license file provided by Carbide 3D, you get access to a script in MeshCam that loads in settings predefined for materials and bits for the Nomad. This makes it very easy for carving into a variety of materials without having to know recommended speeds or any of the other settings. You load your STL, set the max depth and then run the wizard. You then select your material and bit size and everything else is calculated for you.

2D DRAWBACKS

On the PCB milling test, things started to break down. It's not that the Nomad can't handle this; rather, it's the software, or lack thereof, that introduces issues. The tutorials and documentation for the Nomad provide no mention of creating anything other than 3D carvings from STLs. I had to dig through the forums to find possible options for carving PCB designs or other 2D designs. The workflows were complex and I felt like I was jumping through hoops to mill out a PCB or any 2D design.

HEAVENLY HARDWARE, DREAMY DESIGN

While the documentation and design workflow may have left me wanting, the hardware and design of the machine are what dreams are made of. The bamboo sides paired with the all-aluminum frame made it beautiful to look at, and the enclosure meant that I could sit it next to my computer without fear of it being covered in dust. This is a solid machine and because of its sturdy construction, the spindle moves quickly over material without any stuttering. The spindle is made from a brushless motor with a custom speed controller so the cutter can adapt to different materials. The addition of the tool length probe was a very nice touch that made calibrating very easy. The automated calibration paired with an optional clamping fixture upgrade makes 2-sided milling a promising possibility.

The Carbide Motion controller software is very elegant and easy to use. The chatter in the forums also indicates that they are very actively working to improve it and are listening to customer concerns and wishes. It should work with Fusion 360 and a workflow for Eagle to PCB milling is in progress. Carbide 3D has expressed that they are not interested in locking anyone into their workflows and the Carbide Motion software should handle any G-code produced by other CAM tools.

> The hardware and design of the machine are what dreams are made of

CONCLUSION

The absence of thorough documentation and clear workflows is a sign that this is an early product. The focus, however, on hardware design, rigidity, and automated calibration means that an early investment in this product could pay off in the form of downloadable software updates and more support in the near future. Carbide 3D has done a great job of getting the hardware right. It will only be a matter of time before the software catches up. ⊘

MACHINE RATING

⓪①②③④⑤

Ease of Use	
Software Simplicity	
Construction Quality	
Flexibility	
Covet Factor	

PRO TIPS

The included version of MeshCam is a great tool for 3D carvings and makes it easy to transition to CNC if you are used to 3D printing.

You are going to need to find your own software for more traditional 2D carving such as circuit boards.

WHY TO BUY

This thing is solid and it looks great. The aluminum frame keeps the spindle rigid and it moves pretty fast. The enclosure keeps most of the dust and other waste material inside the machine and off of your computer.

SAMPLE CUT

SHAPEOKO 3

WRITTEN BY KURT HAMEL

THIS HEAVY-DUTY MACHINE IS A BIGGER, FASTER, STRONGER VERSION OF THE SHAPEOKO 2

shapeoko.com

MANUFACTURER
Carbide 3D

PRICE AS TESTED
$999

BUILD VOLUME
425×425×75mm

HOST SOFTWARE
Carbide Motion V2

CAM
MeshCam

FIRMWARE
Grbl

OS
Mac, Windows

Available at
Maker Shed

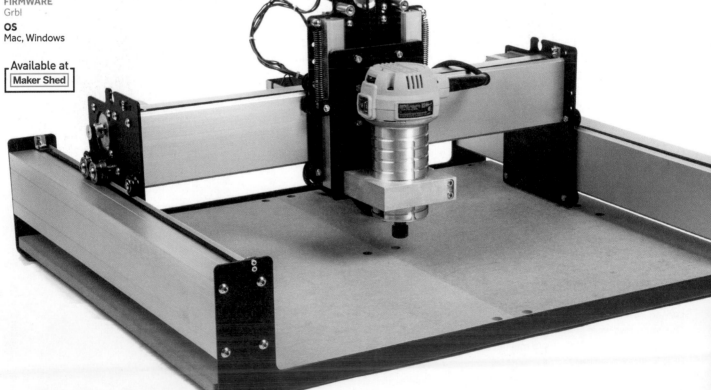

THE SHAPEOKO FAMILY OF CNC CARVERS, ORIGINALLY CREATED BY EDWARD FORD, IS DESIGNED AROUND BEING simple, low cost, and open source. As with the other Shapeokos, the new Shapeoko 3 comes as a kit; our review unit came pre-assembled, but the materials say it can be completed in 2-3 hours.

Ford is producing the Shapeoko 3 in conjunction with Carbide 3D, rather than Inventables, who made the Shapeoko 2. Aside from that, there are only a few major innovations for this iteration, but keeping in line with the Shapeoko mission of CNC simplicity, this makes sense.

RAILS FOR DAYS

The biggest difference from its predecessors and competitors is Shapeoko 3's amazingly thick rails, which provide a super solid frame on which the machine rides. These custom extrusions measure 85mm×55mm — a very sizable increase over the 40mm×20mm rails that came standard on the Shapeoko 2. This provides an incredible amount of strength and rigidity, especially when tackling thick, dense materials.

For those looking to start with a smaller machine but want the option to upgrade later, the Shapeoko 3's rails will come in longer lengths in the future for an easy swap.

THE SPINDLE SAYS IT ALL

With its mount for the DeWalt DWP611 trim router (which you'll have to provide), the Shapeoko 3 is also offering tooling for more hardcore projects than earlier iterations. With it, the Shapeoko 3 looked to be bigger, faster, and stronger than the Shapeoko 2, and a few days of tinkering on this machine confirmed this impression. Additionally, I felt comfortable with the familiar DeWalt installed, much as I did with the Dremel spindle used on the Shapeoko 2 — I didn't need to know much about the electronics, steppers, or gantry system to know what I was dealing with.

INDEPENDENT STUDY

If you know what you're doing on a desktop CNC, making your first cuts on the Shapeoko 3 won't be too much of a hassle. The Carbide 3D Motion software is simple but intuitive. It connected easily to the machine (through a short USB cable) and fed G-code reliably. That's as far up the toolchain as the documentation took me. I was on my own to turn my vector images into G-code. I ended up using a variety of CAM software — Mesh-CAM, MakerCAM, and Easel, all worked fine.

If you are new to desktop CNCs, however, there are a few things that could trip you up. The online documentation assumes the user knows the basic workflow of a CNC tool and is familiar with CAM. The lack of limit switches, which are useful for stopping the carriage from driving itself past the rails' limits, might also be a bother to a beginner, but the open-source lineage of the platform should give you plenty of avenues to add them later should you feel the need. Lastly, you must make your own spoilboard and hold-down clamps. That said, all of these things can be overcome with patience, and I never felt like the required skill level was out of proportion in any one area.

CONCLUSION

The Shapeoko 3 is a great choice if you are finding yourself limited by your Shapeoko 2 or other small CNC carver in terms of power, size, or rigidity. The DeWalt spindle, NEMA 23 stepper motors, deep extrusions, and increased working area are upgrades in all the right areas, and work together to position the machine as a CNC mill and not just a CNC carver.

The Shapeoko 3 is also a great choice for your first CNC mill if you're a Maker whose barrier to entry has been cost. It's a barebones system though, so if you've been holding out until there's something super easy to use, there are better (but costlier) options available. ◐

> Upgrades in all the right areas work together to position the machine as a CNC mill and not just a CNC carver

Gunther Kirsch

MACHINE RATING

0 1 2 3 4 5

- **Ease of Use**
- **Software Simplicity**
- **Construction Quality**
- **Flexibility**
- **Covet Factor**

PRO TIPS

If you want to take advantage of the precision this platform has to offer, ditch the cheap bits that come with the router for some real, precision milling bits. Also consider adding a ⅛" diameter collet to your tool chest so you aren't limited to what the standard ¼" collet will accept.

WHY TO BUY

If you are seeking a low-cost upgrade to your smaller CNC carver in terms of rigidity, power, speed, or work area, the Shapeoko 3 is for you. The simple, open-source nature makes it great for tinkerers, hackers, and anyone looking to learn about what makes a CNC tick.

SAMPLE CUT

OTHERMILL

THIS COMPACT MACHINE IS PERFECT FOR ALL SKILL LEVELS

WRITTEN BY JOSH AJIMA

DESIGN IS CLEARLY A TOP PRIORITY FOR THE COMPACT OTHERMILL, FROM THE GRAPHICS ON THE BOX TO CLEVER TOUCHES such as magnetic side panels and wrench holders milled into the machine. The attractive, white plastic (HDPE) case is enclosed and suitable for home or office, while built-in handles make the Othermill extremely portable.

READY OUT OF THE BOX

Setup was a breeze. Unpacking, inserting the collet, and installing software took only a few minutes. Default settings are configured for a range of materials and I was able to successfully mill circuit boards, machining wax, aluminum, HDPE, and birch plywood.

The Othermill is relatively quiet during operation, only becoming really noticeable when milling aluminum. Removable side panels give access for cleaning out waste materials. Safety features include a fully enclosed design with an automatic motor shut-off when a panel is opened, and an emergency stop button on the side.

SMART SOFTWARE

The Otherplan software, currently Mac-only, guides users through the milling process and provides a smart selection of options based on material selected and file type. The auto Z-homing process between bit changes gives very accurate positioning and is guided by step-by-step on-screen instructions.

Eagle, Gerber, SVG, and G-code files can be directly imported into Otherplan. The tool path visualizations are top-notch and aid in aligning projects on the workpiece. Cutting an SVG file was easy but multiple files are needed for designs involving different cutting depths or nested cutouts.

CONCLUSION

The Othermill is just right for an office, home, design studio, or classroom. Some might find it a little small though, especially when compared to other machines around the same price like the Carvey or Nomad. Simple setup, ease of use, and intuitive software make this an excellent machine for users of all skill levels. ◓

MACHINE RATING

0 1 2 3 4 5

- Ease of Use
- Software Simplicity
- Construction Quality
- Flexibility
- Covet Factor

MANUFACTURER
Other Machine Co.

PRICE AS TESTED
$2,199

BUILD VOLUME
140×114×35.5mm

HOST SOFTWARE
Otherplan

CAM
Otherplan. Can also use third-party software such as Fusion 360 and MeshCAM.

FIRMWARE
Othermill

OS
Mac

othermachine.co

PRO TIPS

Use the included alignment jig for precision or double-sided work. Make your own jig by cutting out an appropriately sized hole in stock material to hold small pieces for milling and engraving. Be sure to leave a cutout to remove the piece from the jig.

WHY TO BUY

The Othermill's compact, quiet, and safe design is perfect for environments where other CNCs might be too large or too loud.

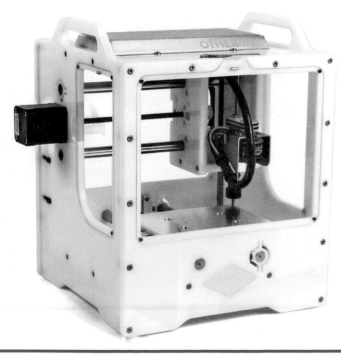

> The Othermill is just right for an office, home, design studio, or classroom

Gunther Kirsch

SAMPLE CUT

MACHINE RATING

0 1 2 3 4 5

Ease of Use	
Software Simplicity	
Construction Quality	
Flexibility	
Covet Factor	

MANUFACTURER
Inventables

PRICE AS TESTED
$1,999

BUILD VOLUME
300×200×70mm

HOST SOFTWARE
Easel

CAM
Easel

FIRMWARE
Custom

OS
Mac, Windows

PRO TIPS

Take what you know and run with it. Using our Eagle and Inkscape knowledge it took us less than 2 hours to come up with a repeatable workflow that was cranking out boards we could be proud of.

Start slow and then speed up. Follow the recommended settings in Easel until you get a handle on how you can push it.

WHY TO BUY

Quiet, clean, and good-looking (with a price point you'd expect for those adjectives), the Carvey can mill out a spot in anyone's tool arsenal.

Inventables

SAMPLE CUT

CARVEY

THIS SMALL, HANDSOME MACHINE QUIETLY PRODUCES NICE WORK WRITTEN BY CHRIS YOHE

THE CARVEY IS A NEW ENTRY INTO THE DESKTOP CNC AND MILLING MARKET FROM INVENTABLES, previously known for their work with Shapeoko and the X-Carve machines. Carvey is a sleek-looking desktop machine, with about the same footprint as a midsize, multifunction printer you often see on desks or tables around the office.

VERSATILE AND QUIET

Once the cover is lifted the inside of the Carvey is easily accessible, giving you a work area slightly larger than a standard sheet of paper, and a depth of 2.75". Carvey comes with a DC spindle that is exceptionally quiet and uses an ER-11 precision Collet which accepts ⅜", ⅛", and ¼" bits. They aren't kidding when they say you can carry on a phone call while it runs — the noise is on par with a desk fan due to the rugged, sealed enclosure.

Inventables touts the machine to work with wood, plastics, waxes, foams, and thin non-ferrous metals as well as PCBs, and we put it through its paces with a little of each. Our tests were done with various ⅛" bits, and even when we pushed the speed we got decent results.

EASING INTO EASEL

Inventables is cleverly banking on Easel, their still-emerging, browser-based carving software. From material selection and size, to milling bit type and measurement, it is all handled through simple, easy-to-understand menus. Helpful hints are given if any missteps seem to have been made.

CONCLUSION

Carvey would be perfect for designers, schools, home users, or your average consumer looking to dip their feet into the water easily. The build size, while one of the smaller we've seen, may be a deterrent for those looking to do larger work such as signs or furniture, but for many users it will suffice for most small to midsized projects. ◑

> The noise of this sleek-looking desktop machine is on par with a desk fan

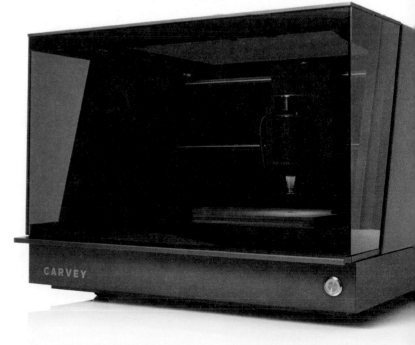

ESSENTIAL END MILLS
FOR CNC MACHINING

YOU'VE GOT YOUR CNC MILL, BUT NOW WHAT? THESE USEFUL BITS WILL GET YOU GOING ON MOST STARTER PROJECTS WRITTEN BY LUIS RODRIGUEZ

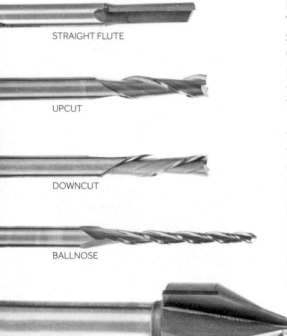

STRAIGHT FLUTE

UPCUT

DOWNCUT

BALLNOSE

V-BIT

COMPRESSION

TABLE-SURFACING

NOW THAT YOU HAVE A CNC MACHINE, YOU NEED TOOLING. BUT DON'T GO STICKING ANY OLD DRILL BIT INTO THE CHUCK. Drill bits are designed to drill, or plunge axially (up and down). For most operations, CNC machines use end mills that cut laterally (side to side).

End mills have cutting surfaces called flutes. The most common end mills have two to four flutes. Generally, fewer flutes evacuate more chips from your material, keeping the bit cool. However, more flutes produce a finer edge finish. There are four basic flute types, each optimized for different materials and edge finish. Solid carbide or carbide tipped are ideal since they don't dull as easily as HSS (high speed steel) end mills.

STRAIGHT FLUTE
These are best for general-use cutting with a good edge quality on many materials.

UPCUT AND DOWNCUT END MILLS
These spiral, flute-shaped end mills either carry chips up and away from the material or down into them. An upcut will keep the bit cool while quickly evacuating materials when cutting plastic or aluminum, however, it will fray the top surface, and may lift your material so be sure to have adequate hold-downs in place. Downcut bits ensure a smooth top surface on laminates, assist with holding down your thin parts, and possibly avoiding tabs on larger parts.

A single "O" flute is key for plastics like HDPE and acrylic when clearing materials. The flute helps avoid excess heat buildup, which may cause material to stick to and ruin the end mill and your part.

BALLNOSE MILL
These bits have a rounded tip and are ideal for 3D tool paths. When combined with a "roughing" bit to clear large areas of material, this end mill will result in smooth 3D surfaces, especially with two or more passes.

V-BIT
A 60° or 90° V-bit is great for what's called V-carving, in which the tip of a V-shaped bit is used to cut into narrow spaces, and the wide bottom is used to cut into larger spaces. V-bits can also create sharp corners that other end mills cannot because of their radiuses.

BONUS END MILLS
COMPRESSION
These bits combine the benefits of both upcut and downcut end mills, ensuring a smooth top and bottom face when cutting laminates and plywood at full depth passes, considerably reducing cut time.

TABLE-SURFACING
These bits are used to surface your table quickly, giving a smooth and level work surface, ensuring accurate cut depths. ✪

Photo by Hep Svadja, End mill cuts by Luis Rodriguez

STRAIGHT FLUTE

UPCUT

DOWNCUT

BALLNOSE

V-BIT

LASER CUTTERS

WRITTEN BY MATT STULTZ AND KACIE HULTGREN

THE HOTTEST TOOL FOR MAKERS IS POWERFUL, VERSATILE, AND SPENDY — HERE'S WHAT YOU NEED TO KNOW TO GET STARTED

Custom lasercut lanterns by **Luminetik Tek**, www.etsy.com/ca/shop/LuminetikTek

IF THERE'S ONE TOOL THAT'S COVETED BY MAKERSPACES MORE THAN ANY OTHER, IT HAS TO BE THE LASER CUTTER. These high-precision machines can produce both functional and beautiful items. Their versatility allows anyone with access to one to quickly go into production with his or her designs.

Go to your local indie craft fair and I bet you'll find laser-cut jewelry. Craft stores are stocked full of laser-cut scrapbooking items. Even big-box stores are not immune to the awesomeness of laser-cut items, offering laser-cut window curtains, holiday ornaments, fixtures, and more.

The power of the laser cutter comes from its ability to cut through a wide range of materials with high precision. Drag-knife cutters — like those on craft cutters and vinyl cutters (see page 76) — can't penetrate hard and thick materials, while a laser can slice through them like butter. And a CNC router has a hard time creating ultra-sharp details (think about cutting a letter V: the outside edges can be cut away sharply with overlapping passes but the inner point can only be as pointy as the diameter of your router bit). A laser's beam is so narrow that it can give you that precise detail.

Building structures with laser-cut parts has become a well-defined practice. Plugins to automatically generate boxes are available for the popular vector graphics application Inkscape, making it a snap to create a case for a project. Captured nuts with tab-and-slot construction make parts easy to assemble and take apart, unlike glued builds. In fact, desktop 3D printing wouldn't be where it is today without laser cutters — MakerBot, Printrbot, SeeMeCNC, Ultimaker, and many other companies started out producing 3D printers largely made from laser-cut parts.

If you're looking for a tool that can widen your boundaries, a laser cutter has a universe of possibilities. Buying your first laser cutter can be a daunting challenge, especially if you're budget-minded. Let this overview of three different types guide you on your way. ◗

LASER CUTTER PROJECTS:

CNC Panel Joinery

How to cut basic cross, tee, and corner joints, snap-together joints, captive-nut joints, and other clever connections on your router or laser cutter. makezine.com/go/cnc-panel-joinery-2

DIY: Luxo Jr. Lamp

A great first laser-cutter project: Build everyone's favorite Pixar mascot out of ¼" and ⅛" acrylic. makezine.com/go/laser-cut-pixar-luxo-lamp

Hep Svadja, Gregory Hayes

BUYING YOUR FIRST LASER CUTTER

WITH GREAT POWER COMES GREAT EXPENSE, BUT HERE ARE THREE ENTRY-LEVEL OPTIONS THAT WILL HAVE YOU BEAMING. WRITTEN BY KACIE HULTGREN AND MATT STULTZ

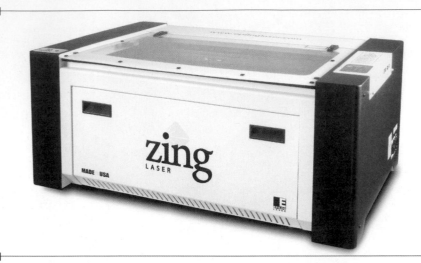

EPILOG LASER

Epilog is the gold standard if you're looking for a high-quality, user-friendly laser cutter. Two big advantages: Epilog's print driver makes the process of sending your designs to the machine quick and painless; and special air-cooled laser tubes made by Epilog reduce the hassles and potential hazards of their liquid-cooled competitors. The convenience doesn't come cheap though; the Zing line, their most affordable, starts at just under $8,000 for a 30-watt, 16"×12" machine.

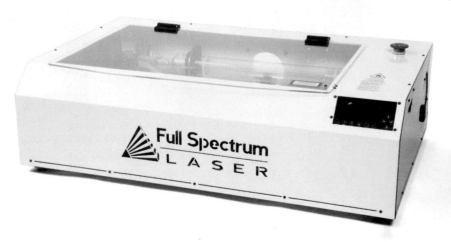

FULL SPECTRUM LASER

Full Spectrum Laser started out by importing Chinese lasers, tuning them up, and rebranding them for American sales. In an attempt to increase reliability and ease of use, FSL decided to create their own laser cutter system and released the H-Series. To reduce cost, the H-Series skips an adjustable bed — instead, you adjust the laser focus by moving the final lens assembly. With a step up in software from the standard import offerings, the H-Series starting at $3,499 offers great value.

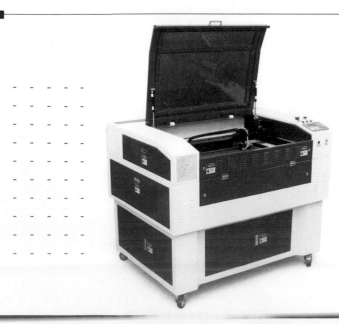

CHINESE IMPORTS

Imported laser cutters are so popular, they're almost a brand themselves. If you're on a budget or looking for the biggest bang for your buck, they are a great option. You might find machines as cheap as $400 on eBay, but expect to pay $1,000 or more for a decent cutter, depending on mods and software upgrades you need to make. Be prepared to swap out small submersible pumps intended to go in buckets for radiant chillers, and to upgrade your ventilation system. The bundled software typically offers a very poor user experience, but that's something you can learn to work around — especially with the potential savings.

Gunther Kirsch

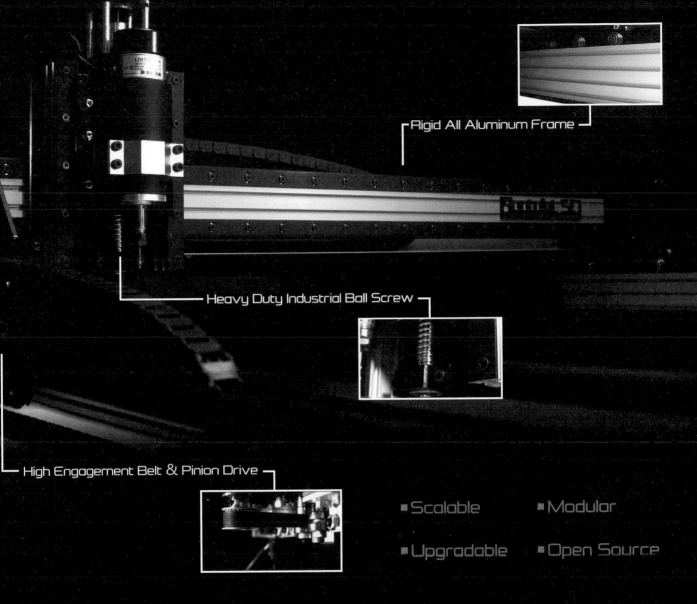

Rigid All Aluminum Frame

Heavy Duty Industrial Ball Screw

High Engagement Belt & Pinion Drive

■Scalable   ■Modular

■Upgradable   ■Open Source

Enter Routakit, a new line of next-gen CNC machines that blur the line between hobby desktop machines and commercial machines, offering best in class performance AND affordability. Etch, carve, and cut a wide variety of materials such as glass, plastic, wood, stone, and metal.

With the open-source files at your disposal, transform your Routakit into a plasma cutter, laser cutter, or even a 3D printer. Choose between either the SD or the HD model to suit your needs and budget.

What Will You Make?

www.routakit.com

WHY YOU NEED A VINYL CUTTER

WELCOME TO THE WORLD OF CUTTING PLOTTERS, YOUR NEWEST DIGIFAB ADDICTION (EVEN IF YOU DON'T KNOW IT YET)

WRITTEN BY MATT GRIFFIN

My Other Computer Is a Robot

OF ALL THE DIGITAL FABRICATION TOOLS EXPLORED IN THIS ISSUE, cutting plotters have spread the furthest into the consumer marketplace, thanks in part to the evangelism of Oprah, Martha Stewart, and thousands of Etsy crafters worldwide.

These are no longer just for Christmas design patterns you can buy from retail outlets. At the heart of these plastic shells lies a computer-controlled craft robot that drives a selected tool (often a blade) along paths set by the vector-graphic design file submitted to the device's control software. If you notice a similarity to some 3D printers, it's no accident: Cutters played a crucial role in inspiring the designers of desktop extrusion-based 3D printers.

There are many names for these tools, varying based on the device's intended use, but "cutting plotter" is the most broad, and "vinyl cutter" is one of the most common. Don't pick a machine just based on speed, force, and price — look at drive motor, tracking, form factor (craft cutter or roll-feed plotter), and your intended use. The three machines you'll see here are all useful for projects from stickers and labels to foamcore prototyping and building PCB circuits. So skim past the "Buy Here" links for holiday trifles, flip open the hood of the cutter, and consider the diverse range of uses at your fingertips. ◉

VINYL STICKER PROJECTS:

Make:

Silk-Screen Printing with Vinyl Stencils

Make your own custom silk-screened T-shirts and posters — a great first vinyl cutter project. makezine.com/go/vinyl-silk-screen-printing

Vinyl Cut PCB Stencils

Etch copper circuit boards the easy way — with a perfect "resist" pattern you cut in vinyl. makezine.com/go/vinyl-cut-pcb-resist

SILHOUETTE CAMEO

A LEADING FAVORITE OF PAPER CIRCUIT AND PAPERCRAFT MAKERS

WRITTEN BY MATT GRIFFIN

SILHOUETTE'S FLAGSHIP CAMEO SERIES HAS SECURED A STRONG FOLLOWING among Makers, papercrafters, and fab labs since it first launched in 2011. The current version cuts dependably and is easy to use, while adding nice features like touchscreen, automatic trimming capability, and an optional roll feeder that allows (much) longer prints.

NOT FOR CRAFTERS ONLY

Many craft cutter brands target scrapbook, letterpress, and home-craft customers. While the Silhouette Store does have a large and diverse premade design catalog, their Designer Edition software upgrade also allows operators to run their own designs — and their own wild ideas — earning a Maker-friendly reputation thanks to impressive projects by community members, from PCB etching to custom origami.

Although the Cameo hasn't changed fundamentally since it first appeared, the 2014 update introduced a number of helpful new features. A large digital touchscreen replaced a set of hardware buttons, making it far easier to cut jobs and align complex projects while untethered. A handy parts compartment stores spare blades (which you should stock up on, as they are proprietary and wear out quickly), and firmware updates added new UI features.

STAY IN LINE

At the back of the Cameo, a slot for a crosscutting tool allows rolls of vinyl to be trimmed straight across (best suited for use with the vinyl roller add-on). Inside the machine itself, steel spring rollers maintain accurate feeds, and alignment marks help with matted and non-matted feeding.

Scrapbookers will love the preprinted 12"×12" standard-size loose sheet sets. The size opens up the printer to 12" rolls of vinyl, A3 and A4 papers, and grants even more territory for 3D Pepakura and kirigami paper craft projects.

CONCLUSION

For everything from cute greeting cards to sophisticated PCB etching and beyond, the Cameo is full of features that will make getting started easy, but keep experts coming back. ◗

MANUFACTURER Silhouette America

PRICE AS TESTED $299

BUILD AREA Flat: 305×305mm; Roll: 305×3,048mm

MATERIAL THICKNESS 31.5 mils (0.8mm)

TOOL PRESSURE 210 gram force

BED STYLE Materials fed via a roll or on a cutting mat

MATERIALS Vinyl, heat transfer material, cardstock, photo paper, copy paper, rhinestone template material, fabric, and more

BLADE TYPE Proprietary (fabric blade available)

PEN TYPE Proprietary (pen holder available separately)

OPERATE UNTETHERED Yes (USB stick)

ONBOARD CONTROLS Digital touchscreen for monitoring and execution on the device

HOST SOFTWARE Silhouette Studio

OS Mac, Windows

OPTIONAL ACCESSORIES Numerous, listed on Silhouette's site

PRO TIPS

To prevent super sticky Silhouette medium-hold cutting mats from tearing away the bottom layer of sensitive materials, first press a soft, worn pillowcase or sheet to the page.

The Cameo's feed mechanism can cut sticker-backed vinyl sheets without a cutting mat or carrier sheet.

Upgrading from Silhouette Studio to the Designer Edition or a third-party tool such as Sure Cuts A Lot Pro may be required for some high-detail custom cuts.

WHY TO BUY

Fast, accurate, and versatile, the Cameo is one of the few desktop craft cutters that can reach every part of a 12"×12" sheet. The addition of the roll feeder allows the device to print long, narrow designs up to 10' in length.

Gunther Kirsch

Their Designer Edition software upgrade allows operators to run their own designs — and their own wild ideas

silhouetteamerica.com

MH871-MK2 WRITTEN BY MATT GRIFFIN

A STANDING VINYL CUTTER FOR THE PRICE OF A DESKTOP VERSION

MANY ROLL-FEEDING CUTTERS COST MORE THAN MAKERS WILL WANT TO SPEND FOR SUCH A MATERIAL-SPECIFIC TOOL. They're priced to be workhorses for professional sign shops. But if you're willing to fiddle and fuss with USCutter's MH series, you can get clean cuts and basic functionality for thousands less in a more professional form factor than the craft cutters we reviewed. These discount cutters require much more attention and care than the premium versions. "You get what you pay for" applies, but so does "if you can fix it, you truly own it."

Officially, this cutter is PC only, but we were able to make it work on a Mac with third party USB to serial hardware and SCAL Pro.

CONCLUSION

Makers willing to roll up their sleeves to get to know the machine and master the fussy pinch rollers may find that they have the best of both worlds: A hobby cutter that can cut small things nearly as well as a dedicated desktop craft cutter, but still cuts large things nearly as well as an expensive sign shop vinyl cutter. ⊘

MANUFACTURER USCutter
PRICE AS TESTED $290
BED AREA 780mm wide×roll length
MATERIAL THICKNESS 1mm
TOOL PRESSURE 600 gram force
BED STYLE Roller fed
MATERIALS Vinyl, cardstock, paint mask, heat transfer
BLADE TYPE Open, carriage has 1 blade slot and 1 plotter slot
PEN TYPE Holds ballpoint pens (adapter for other pens available)
OPERATE UNTETHERED? No, USB required
ONBOARD CONTROLS? Yes, can adjust speed, pressure, set calibration points
HOST SOFTWARE Sure Cuts A Lot Pro
OS Windows only
OPTIONAL ACCESSORIES Floor Stand

PRO TIPS

To avoid wrinkling, buckling, and tearing, adjust the tension and placement for the pinch-rollers.

Experiment with the Cut Mode setting to begin your cut from the perfect location.

The quality — and type — of vinyl you use has a huge effect on how the machine handles your print.

WHY TO BUY

At a price similar to desktop craft cutters, this hardy, hobby-grade vinyl cutter offers longer cuts on thicker material out of the box.

uscutter.com

PORTRAIT WRITTEN BY MATT GRIFFIN

A PINT-SIZED PLOTTER FOR A REASONABLE PRICE

THE EASY-TO-RUN PORTRAIT OFFERS MANY OF THE PREMIUM CAPABILITIES OF THE CAMEO (see page 77) while lowering costs by forgoing the LCD display and USB stick slot. It's smaller too, but with the optional roll feeder, can cut work as long as the Cameo, albeit not as wide. It cuts the same variety of materials with the same precision and tool pressure, in an impressive, toss-it-in-your-bag, featherweight 3½-pound package. However, the cutting area, limited to a little less than a letter-size page, can be a serious limitation, and the lack of untethered cutting can make it harder for teams and classes to share a single Portrait.

CONCLUSION

For Makers working small or building larger projects piece by piece, the cutting area will accomplish many of your tasks, from laptop decals to temporary tattoos to painting stencils. By using Silhouette Studio Designer Edition or third-party software like SCAL Pro or Make the Cut, this affordable machine can accurately cut original designs, unlike competitor's entry-level plotters that largely require purchasing existing designs ⊘

MANUFACTURER Silhouette America
PRICE AS TESTED $199
BED AREA Flat: 203×305mm; Roll: 203×3,048mm
MATERIAL THICKNESS 31.5 mils (0.8mm)
TOOL PRESSURE 210 gram force
BED STYLE Cutting mat or roller fed
MATERIALS Same as Silhouette Cameo
BLADE TYPE Proprietary (fabric blade available)
PEN TYPE Proprietary (pen holder available)
OPERATE UNTETHERED? No (USB required)
ONBOARD CONTROLS? Engage and eject cutting mats and materials
HOST SOFTWARE Silhouette Studio
OS Mac, Windows
OPTIONAL ACCESSORIES Numerous, listed on Silhouette's site

PRO TIPS

Upgrade to Designer Edition or a third-party tool for more file-type options, offering a wider variety of software and better control of the Portrait's smaller cutting area.

Narrow designs such as signs and banner elements can be run out to 10 feet (finesse required!)

WHY TO BUY

Those looking for a pint-size cutting plotter they can bring with them (or stow away when not in use) will appreciate the lower price, while retaining most of the capabilities of the Cameo.

silhouetteamerica.com

Heather Kirsch

ONES TO WATCH

THESE SLEEK MACHINES ARE POISED TO CHANGE 3D PRINTING   WRITTEN BY CALEB KRAFT

PRINTING WITH PLAIN PLASTIC JUST ISN'T EXCITING ANYMORE. The big things coming on the horizon all deal with new and interesting materials or mixes of materials. Some of these add functionality directly into the printed piece, while others use altogether new processes — we're excited for the yet undelivered desktop powder sinterers. Hopefully we'll be seeing these machines on our bench soon.

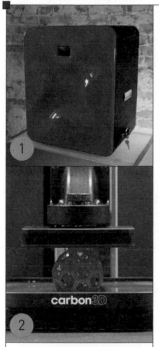

SINTERIT ❶ is bringing SLS printing to the desktop, where lasers burn fine powders into physical objects. At $5,000 or less, these printers will appeal to those who have a big wallet and need better quality than standard FDM printing.

CARBON3D'S ❷ resin-based printer made waves this year with their "continuous printing" technology that allows an item to be created very quickly with almost no visible layers. The extreme speed at which this system uses UV light to solidify resin — stated to be an incredible 25 to 100 times faster than current SLA printers — will change the use of printers in manufacturing.

Typical prints from almost any 3D printer are not quite strong enough for industrial use, especially if what you need is both strong and light. **MARKFORGED** ❸ is embedding their prints with carbon composites, allowing for fully 3D printed parts that can withstand pressures similar to metal. Their pitch is that you can now print functional parts that used to require a mill to fabricate.

VOXEL8 ❹ wants to expand the function of the 3D printer outside of passive parts. By printing in conductive materials you can embed circuits directly into your prints. Simply add the electronic components and you're finished. Printing entire functional electronic gizmos isn't too far away for this group.

ACCESSORIES   WRITTEN BY MATT STULTZ

So what do you get for the 3D printing enthusiast in your life when they already have a printer (or two, or five)? How about some great accessories!

DISCOV3RY PASTE EXTRUDER
$379
STRUCTUR3D.IO/DISCOV3RY-PRODUCTS/DISCOV3RY
The more materials you can extrude out of your 3D printer, the more useful it becomes. The Discov3ry Extruder by Structur3D makes printing pastes easy. Just move your extruder motor's connection to the Discov3ry's motor, and place the nozzle by your hot end nozzle. Any substance about the consistency of creamy peanut butter can be extruded — this includes materials like silicone, thinned clays, or even Nutella!

MATTERCONTROL TOUCH TABLET
$299
MATTERHACKERS.COM/STORE/PRINTER-ACCESSORIES/MATTERCONTROL-TOUCH
There are many ways to control your 3D printer: USB to your computer, on-board controls, or Raspberry Pi are just a few. The MatterControl Touch is an Android tablet built for running your 3D printer. It is preloaded with the MatterControl software that allows you to not only control the printer from the tablet, but also to slice your STL files. Files can be downloaded over Wi-Fi from popular file sharing sites, or synced with services like Dropbox or Google Drive. This is my go-to solution for running printers on the go.

PRINTINZ ZEBRA PLATE
STARTING AT $13
PRINTINZ.COM/ZEBRA-PLATES
Keeping your prints stuck on your build plate is important, but so is getting them off when they're complete. The Zebra Plate replaces your glass or acrylic plate and has a great surface for prints to stick to. When it comes time to remove them, simply flex your plate to pop them off. The Zebra Plate has an inner spring-steel lining to help keep your leveling sensors working, but also to allow the plate to flex and return to its original shape.

PROTO PASTA WEAR RESISTANT NOZZLE
$14.99
PROTO-PASTA.COM/COLLECTIONS/RETAIL/PRODUCTS/PLATED-BRASS-WEAR-RESISTANT-NOZZLES
There are lots of great filaments coming to market with other materials added to the plastics to provide new properties. Many of these — like carbon fiber, steel, bronze, or other metals — are very abrasive to your printer's nozzle. You can print with confidence using the wear resistant nozzles from Proto Pasta.

MANUFACTURERS, LISTEN UP:

We asked our expert reviewers what 3D printer innovations they hope become commonplace. Here's what they said.

Josh Ajima— "Detection of filament diameter, extrusion width, extrusion volume, and gaps"

Spencer Zawasky, Claudia Ng— "Filament monitors and feedback for clog detection"

Matt Griffin— "Printers as network devices running on 3D host apps rather than tethered to laptops"

Kurt Hamel— "Wi-Fi and Bluetooth standard"

Chris Yohe— "Automatic web progress updates"

Samuel Bernier— "Better dissolvable support material"

Chandi Campbell— "More accessible, quick-change style extruder assemblies and/or nozzles"

Chris Yohe— "Temperature runaway shutoffs"

Shawn Grimes— "Auto bed leveling"

"Zortrax provides solutions for professionals who need 3 things: quality, an easy learning curve, and a competitive price. Zortrax's new product – the Inventure 3D printer – has them all."

3D printing technology is now a fixture in the workspace of designers and engineers. However, many entrepreneurs are still afraid that quick in-house prototyping is expensive and time-consuming. Professionals don't like compromise and expect proven solutions at competitive prices.

The answer is the Zortrax Inventure – a new solution from the maker of the proven Zortrax M200, which has thousands of satisfied users all over the world.

The Right Quality for Industrial Use

Thanks to the introduction of **a closed heated chamber** and soluble support material, **Zortrax Inventure** is capable of printing complex, detailed models with numerous moving parts, durable prototypes and any complex objects which would be difficult to print with less advanced desktop 3D printers.

Owing to its constant temperature, the heated chamber ensures the **high dimensional accuracy, smooth surface and precise details** of the printed 3D models. When combined with the Dissolvable Support System, Zortrax Inventure can create most advanced 3D objects, such as architecture mockups, moving parts, mechanical elements and advanced prototypes – all printed in one piece, with an easily removable support structure which leaves no marks on the model.

Z-ULTRAT Plus - durable, thermoplastic material dedicated for Inventure

Mechanical prototype, 3D printed in one element with Zortrax Inventure

Inventure Can Print even the Most Complicated Models

Thanks to the double nozzle, the device can print an object using two materials – the support and the base – simultaneously, with the former completely dissolvable in water. As a result, the printed models do not require manual removal of the support, which allows for the creation of more complex objects, such as prototypes with movable elements requiring support within the structure, or objects with a complex interior.

Zortrax is Proud of its Quick and Easy Learning Curve

When choosing **Zortrax**, you are buying more than just a 3D printer. Zortrax is a complete ecosystem, with the printing device comprising just one of its elements. The product includes the original **Z-SUITE** software equipped with an internal library of ready-to-use models, allowing you to store all of your projects in one place, improve them and share both your projects and your experiences. Z-SUITE has many advanced features which make the process quick and intuitive.

The Arrival of the New Z-ULTRAT Plus Material

Z-ULTRAT Plus is an exceptionally durable thermoplastic material, perfect for prototyping working elements and printing end use parts. Due to its ability to reflect the properties of objects manufactured in injection molding technology, it can be used for accurate pre-production testing.

The Zortrax Ecosystem

The integrated **Zortrax Ecosystem, which consists of the 3D printer, software and dedicated materials**, provides top quality prints without unnecessary testing or filament selection. This not only allows companies to save time and money but, most importantly, to **ensure the same high quality of repeated printouts**.

New Opportunities for the Business Sector

Zortrax Inventure is an attractive solution for a whole range of industries. Combined with the **Z-ULTRAT Plus** material, this professional, compact 3D printer is capable of creating precise, detailed prototypes which are ready for iteration and testing during the product development process. **Zortrax Inventure** at your office or workshop will not only lower prototyping costs, but will noticeably accelerate and facilitate the process. With 3D technology at the touch of a button, you can make multiple prototypes quickly and at low cost, which in turn allows you to perform numerous tests to improve your products. The result? Your company's competitive edge on the market.

 *Model with support* ....>  *Model without support*

Dissolvable Support System allows you to create support structure which is simply soluble in water.

Read more about 3D printing for business at www.zortrax.com

3D FABRICATOR QUICK GUIDE

THE BEST MACHINES FROM *MAKE:'*S
TESTS AND REVIEWS

THE BEST FDM 3D PRINTERS

FOR THE DESIGNER, TEACHER, TINKERER, AND ENGINEER, THESE ARE THE STANDOUT MACHINES IN EACH CLASS

Best OVERALL
The machines with the best total score

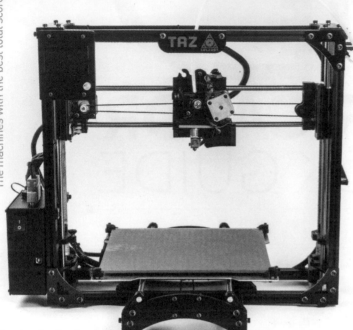

2ND:
ZORTRAX
M200

34
TOTAL SCORE

If you care about 3D prints more than the process of 3D printing, you need to look at the Zortrax M200.

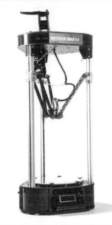

3RD:
ROSTOCK MAX

33
TOTAL SCORE

Makes huge and beautiful prints. You won't break the bank with the Rostock Max.

1ST: TAZ 5

35
TOTAL SCORE

The fifth version of the Taz shows LulzBot's commitment to excellent engineering.

Best VALUE
Greatest cost-to-features ratio

1ST: ROSTOCK MAX
Its large print volume combined with the cost savings of a kit makes this machine a sure bet.

2ND: PRINTRBOT SIMPLE
Expandability at a low price. This is one of the top machines for getting started in 3D printing.

3RD: PRINTRBOT PLAY
Great prints for just $399. Need we say more?

Gunther Kirsch, Kelly Egan

Best for SCHOOLS
Safety features and ease of use

1ST: PRINTRBOT PLAY
Its low cost and safety oriented elements make the Play a good machine for students and teachers.

2ND: UP BOX
Fully enclosed, good safety features, and ease of use should make this a classroom hit.

3RD: BEEINSCHOOL
This sturdy machine designed for schools offers educational pricing.

Most PORTABLE
Mobile or space-saving machines

1ST: SIMPLE
Still one of the best starter printers. And now with the included handle, a great mobile machine.

2ND: ULTIMAKER GO
The shipping foam doubles as a carrying case to take the Go on the go.

3RD: PRINTRBOT PLAY
Small and light, the Play is easy to transport.

Outstanding OPEN SOURCE
The beginning and future of 3D printing

1ST: TAZ 5
LulzBot keeps striving to make the Taz line better while still holding true to its open source roots.

2ND: ROSTOCK MAX
Huge print area, great prints, and open source — it's a 3D printer trifecta.

3RD: ULTIMAKER GO/EXTENDED
Ultimaker brings design and beauty to a machine that you could still largely build on your own.

Best LARGE FORMAT
Sizeable machines for those who want the biggest prints

1ST: ULTIMAKER EXTENDED
The Ultimaker Extended gives you a great print area while not taking up your entire desk.

2ND: ROSTOCK MAX
If you want to go tall, this is the machine for you.

3RD: TAZ 5
Big beautiful prints — if you have the desk space.

NOTABLE CNC MILLS
YOUR BEST BETS FOR CUTTING AND ROUTING PROJECTS, BIG, MEDIUM, OR SMALL

Best LARGE FORMAT
Big machines ready to cut your next piece of furniture

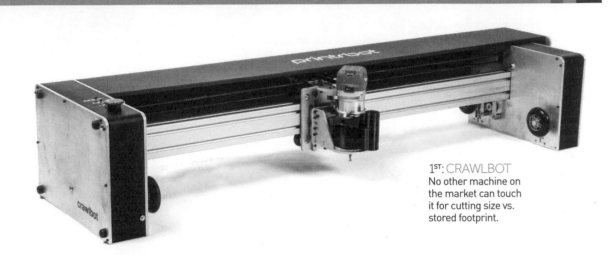

1ST: CRAWLBOT
No other machine on the market can touch it for cutting size vs. stored footprint.

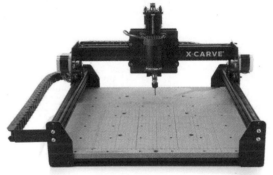

2ND: X-CARVE 1,000MM
This is a nice-sized machine if you don't have the room to cut full sheets.

3RD: SHAPEOKO 3
Decent size out of the box that can be expanded easily thanks to its heavy-duty rails.

Best MID-SIZE
Great hobby mills that can make tons of projects and get you into the CNC world

1ST: SHOPBOT DESKTOP
A well built and easy to use workhorse. But you pay for it.

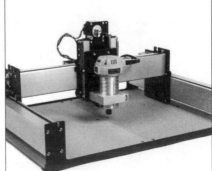

2ND: SHAPEOKO 3
Simple and rigid, exactly what you want in a CNC mill.

3RD: PRINTRBOT CNC
The Printrbot team threw out the CNC-making playbook to bring us this Crawlbot.

Best DESKTOP MILL
PCBs, molds, and small parts are a click away with these desktop workhorses

1ST: NOMAD 883
It may look pretty with its bamboo case, but the fantastic clamping system makes this a machine for serious use.

2ND: CARVEY
Quiet and easy to use, perfect for a desktop mill.

3RD: OTHERMILL
When you want a highly portable mill, the Othermill is your best option.

SLA Printer
SELECTIONS
WITH A GROWING NUMBER OF CHOICES, TWO MACHINES STAND TALL IN THIS FIELD

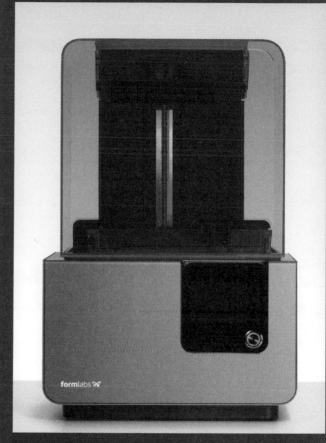

formlabs

1ST: FORM 2
A large print area, auto-fill resin, open resin compatibility, and Wi-Fi connectivity — just a few things that put the Form 2 on top of the resin printer market.

2ND: LITTLERP
If you are interested in getting started with resin printers, the LittleRP is a low-cost, easy-to-build kit. Toss in your own DLP home theater projector and you are off and printing.

BY THE NUMBERS

TEST SCORES AND MACHINE SPECS, SIDE BY SIDE

There's no one machine that can do everything for everyone. Use these charts to help find the right digital desktop fabricator for your needs. (Scores based on criteria listed in *Make:* Volume 48.)

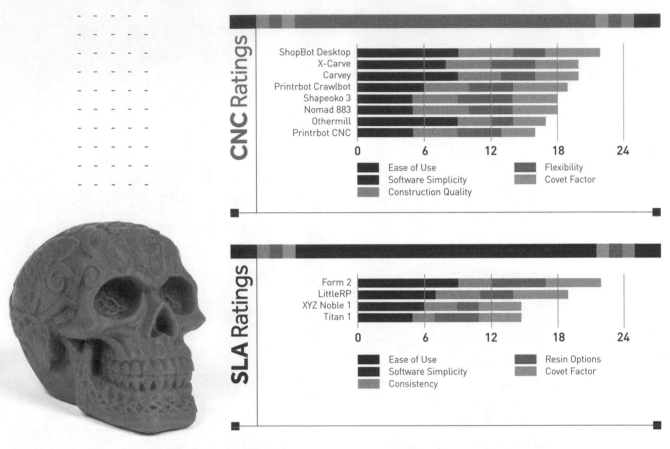

CNC Ratings

ShopBot Desktop	
X-Carve	
Carvey	
Printrbot Crawlbot	
Shapeoko 3	
Nomad 883	
Othermill	
Printrbot CNC	

0 6 12 18 24

- Ease of Use
- Software Simplicity
- Construction Quality
- Flexibility
- Covet Factor

SLA Ratings

Form 2	
LittleRP	
XYZ Noble 1	
Titan 1	

0 6 12 18 24

- Ease of Use
- Software Simplicity
- Consistency
- Resin Options
- Covet Factor

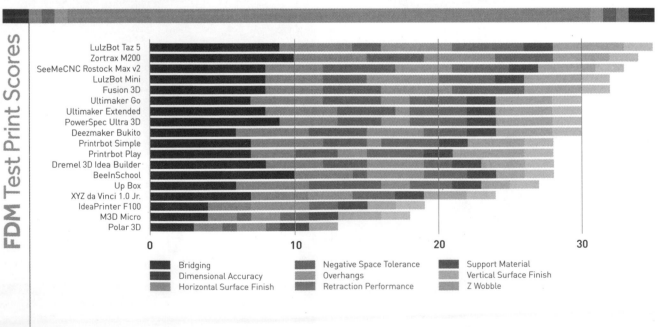

FDM Test Print Scores

LulzBot Taz 5	
Zortrax M200	
SeeMeCNC Rostock Max v2	
LulzBot Mini	
Fusion 3D	
Ultimaker Go	
Ultimaker Extended	
PowerSpec Ultra 3D	
Deezmaker Bukito	
Printrbot Simple	
Printrbot Play	
Dremel 3D Idea Builder	
BeeInSchool	
Up Box	
XYZ da Vinci 1.0 Jr.	
IdeaPrinter F100	
M3D Micro	
Polar 3D	

0 10 20 30

- Bridging
- Dimensional Accuracy
- Horizontal Surface Finish
- Negative Space Tolerance
- Overhangs
- Retraction Performance
- Support Material
- Vertical Surface Finish
- Z Wobble

Hep Svadja

FDM Comparisons

Machine	Taz 5	M200	Rostock Max v2	LulzBot Mini	F306	Ultimaker 2 Go	Ultimaker 2 Extended	PowerSpec Ultra	Bukito
Manufacturer	LulzBot	Zortrax	SeeMeCNC	LulzBot	Fusion3	Ultimaker	Ultimaker	Micro Center	Deezmaker
Cost	$2,200	$2,000	$999	$1,350	$3,975	$1,335	$2,788	$799	$849
Build Volume	298×275×250mm	200×200×180mm	280mm dia.× 375mm	152×152×158mm	306×306×306mm	120×120×115mm	223×223×305mm	229×150×150mm	125×150×125mm
Open Filament	Yes	No	Yes	Yes	Yes	Yes	Yes	Yes	Yes
Bed Style	Heated Glass w/ PEI	Heated Perf board	Heated Glass	Heated Glass with PEI	Heated Mirrored Glass	Unheated Glass	Heated Glass	Heated Plastic	Unheated Acrylic
Print Untethered	Yes	Yes	Yes	No	Yes	Yes	Yes	Yes	Yes
Open Source	Yes	No	Yes	Yes	No	Yes	Yes	No	No
Total Score	35	34	33	32	32	30	30	30	30

Machine	Simple	Play	3D Idea Builder	BeeInSchool	Up Box	da Vinci 1.0 Jr.	IdeaPrinter F100	Micro	Polar 3D
Manufacturer	Printrbot	Printrbot	Dremel	BeeVeryCreative	3D Printing Systems	XYZprinting	Fusion Tech	M3D	Polar 3D
Cost	$599	$399	$999	$1,647	$1,899	$349	$1,200	$349	$799
Build Volume	150×150×150mm	100×100×130mm	230×150×140mm	190×135×125mm	255×255×205mm	150×150×150mm	305×205×175mm	109×113×116mm	203mm dia.× 152mm
Open Filament	Yes	Yes	No	Yes	No	No	Yes	Yes	Yes
Bed Style	Unheated Aluminum	Unheated Aluminum	Unheated BuildTak	Unheated Acrylic	Heated Perf Board	Unheated Glass w/ Custom Tape	Unheated Acrylic	Unheated BuildTak	Unheated Mirrored Glass
Print Untethered	Yes	Yes	Yes	Yes, after print has started	Yes	Yes	Yes	No	Yes
Open Source	Yes	Yes	No	No	No	No	No	No	No
Total Score	28	28	28	28	27	24	19	18	13

SLA Comparisons

Machine	Form 2	LittleRP	Nobel 1.0	Titan 1
Manufacturer	Formlabs	LittleRP	XYZprinting	Kudo3D
Cost	$3,499	$599	$1,499	$3,208
Build Volume	145×145×175mm	60×40×100mm	128×128×200mm	192×108×243mm
Style	SLA	DLP	SLA	DLP
Open Resin	Yes	Yes	No	Yes
Print Untethered	Yes	No	Yes	No

CNC Comparisons

Machine	ShopBot Desktop	X-Carve	Carvey	Crawlbot	Shapeoko3	Nomad 883	Othermill	Printrbot CNC
Manufacturer	ShopBot	Inventables	Inventables	Printrbot	Carbide 3D	Carbide 3D	Other Machine Co.	Printrbot
Cost	$7,330	$1,157	$1,999	$3,999	$999	$2,599	$2,199	$1,499
Build Volume	610×457×140mm	300×300×65mm	300×300× 70mm	48×96×2in	425×425×75mm	203×203×76mm	140×114×35.5mm	457×355×101mm
Cam	VCarve Pro	Easel	Easel	Fusion360 or Printrbot Software	MeshCam	MeshCam	Otherplan	Fusion360

UP BOX

Big
news.

Introducing the UP BOX for professionals.

With a large build volume, high resolution output, and 30% faster print speed, the UP BOX is the ultimate 3D printer for prototyping.

| BUILD VOLUME | MICRON PRINTING | FULL PRINTING |
| 8" 8" 10" | 100μ | ABS PLA |

New website.

A fresh new look. But the same great products you've come to trust. From the affordable UP mini, to the versatile UP Plus 2, to our new flagship UP BOX, drop in and see build quality.

up3dusa.com | **up3d.com**
(FOR USA CUSTOMERS)

support@pp3dp.com   1-888-288-6124

DESIGNED AND MANUFACTURED BY TIERTIME.

UP
Build. Quality.

Home Automation, Robots and Maker Projects built with Windows 10 IoT Core.
We have the perfect Starter Kit with Raspberry Pi 2 for you.

Order your kit today from:

hackster.io/windows10kit

Visit windowsondevices.com for instructions to get started with this kit.

GIFT GUIDE

THE BEST TOOLS AND TOYS FOR THE ONES YOU LOVE

We scoured the world for some of our favorite products — gifts for Makers of every type, from the best solder sucker to a speedy FPV microdrone. Go online to find even more ideas and categories: makezine.com/go/2015-gift-guide.

3D ROBOTICS SOLO
$1,400: 3DROBOTICS.COM
Smart as a whip with its dual Linux processors, the latest quadcopter by 3D Robotics is designed to film the action with minimal input from the pilot. Its gimbal includes the first ever interface with the GoPro bus, allowing camera control from your transmitter (camera not included; see below).

DJI PHANTOM 3 PRO
$1,259: DJI.COM
Compact and stable, the Phantom 3 Pro puts a high-quality camera package into a capable aerial platform. This relative affordability combined with astounding ease of use has helped Phantom quadcopters become the most commonly seen rigs in the air.

Hep Svadja

HOVERSHIP 3DFLY FPV MICRO DRONE KIT
$110: MAKERSHED.COM
Started as a full-scale 3D printable racing quadcopter, Hovership has since launched a variety of products geared towards the FPV enthusiast. The new 3DFly is a micro racer perfect for quick trips around the house or park.

54 PIECE BIT DRIVER SET
$25: MAKERSHED.COM
Face it, if you're into flying drones, you're going to have to do a lot of repairs. This comprehensive set of precision bits will be a lifesaver.

GOPRO HERO4 BLACK
$500: GOPRO.COM
While some drone makers are now incorporating their own cameras, the GoPro Hero4 Black remains the leader in action cameras, with the highest quality video in a compact, lightweight form. GoPro's new Session camera ($300) is smaller, but the Hero remains the best rig for capturing your flight.

TENERGY TB-6B 50W BALANCING CHARGER
$40: TENERGY.COM
Over the past few years, the introduction of LiPo batteries has helped aerial hobbyists get airborne with their high power-to-weight ratios. They're a little bit more finicky to charge though. Avoid unwanted overcharging and keep your cells balanced with the TB-6B.

Written by Mike Senese

DRONES

makezine.com/go/gift-guide-2015

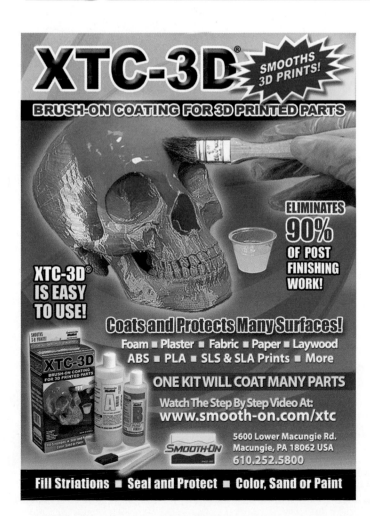

2015 Holiday

GIFT GUIDE

THE BEST TOOLS AND TOYS FOR THE ONES YOU LOVE

WESTCOTT TITANIUM BONDED, NON-STICK SCISSORS

$8–$18: WESTCOTTBRAND.COM

Not all scissors are created equal. While it's often recommended to get a different pair of scissors for each material you're going to be cutting regularly, that's not always feasible. If you're going to get just one pair, get some titanium bonded nonstick scissors. They'll stay sharp and clean.

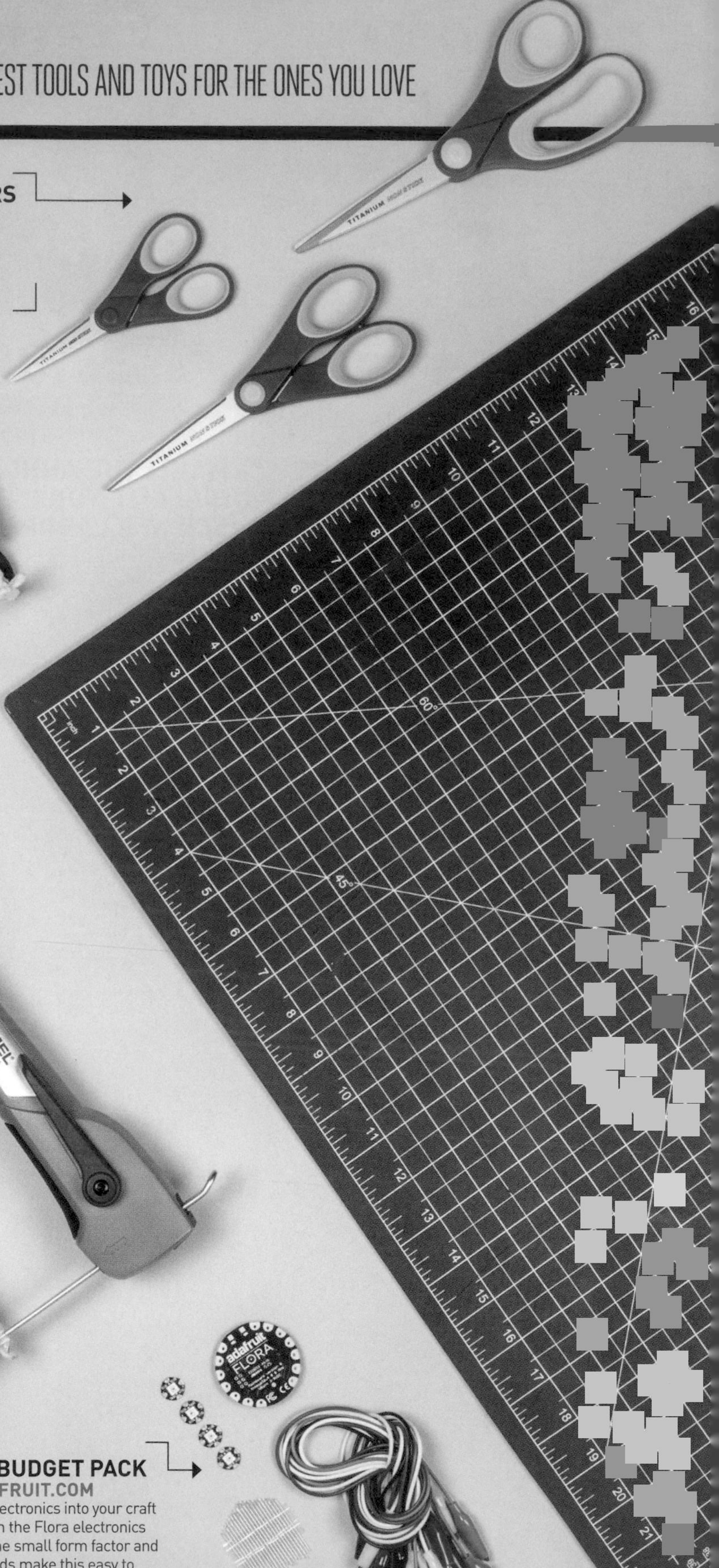

HAND STITCHED OWL KIT

$20: MAKERSHED.COM

New to sewing? Hone your skills on this adorable owl kit. All of the parts you need are included in the box, as well as easy to follow instructions. Use this cute hooter as a springboard into your stuffed animal-making hobby.

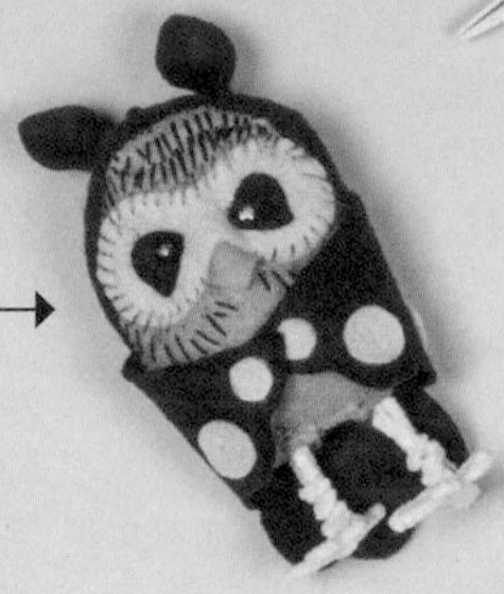

DAHLE SELF-HEALING CUTTING MAT

$12–$120, DEPENDING ON SIZE: DAHLE.COM

Your work surface can make a project easy or difficult. A self-healing cutting mat is fantastic to work on, even when you're not cutting. Built-in grid lines and a nice feel keep you focused on your project, and the surface won't dull your knives the way others can.

DREMEL MOTO-SAW

$75: DREMEL.COM

Dremel is the quintessential Maker's tool, with unmatched versatility. The Moto-Saw attachment brings even more possibilities to your desktop, whether you're cutting shapes from wood, fiberboard, or even thin metal.

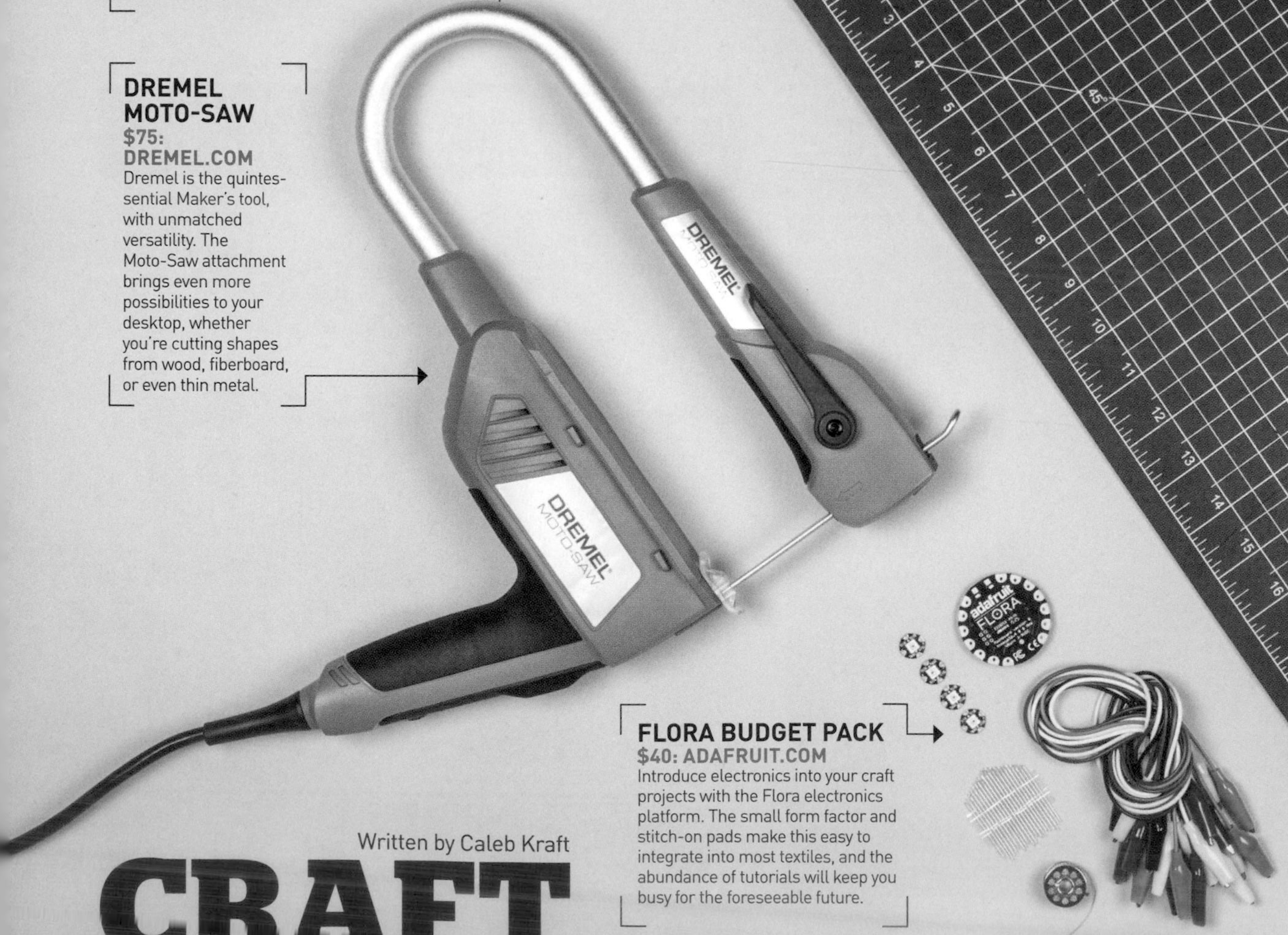

FLORA BUDGET PACK

$40: ADAFRUIT.COM

Introduce electronics into your craft projects with the Flora electronics platform. The small form factor and stitch-on pads make this easy to integrate into most textiles, and the abundance of tutorials will keep you busy for the foreseeable future.

Hep Svadja

Written by Caleb Kraft

CRAFT

makezine.com/go/gift-guide-2015

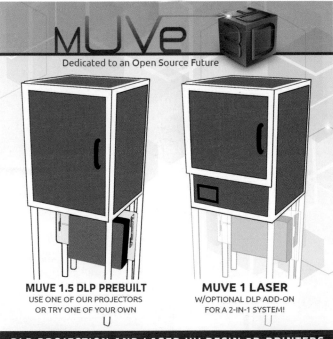

SCULPTURE DEPOT ADJUSTABLE ARMATURE STAND
$64–$77: SCULPTUREDEPOT.NET
A good sculpting experience starts with a good armature stand, and nothing is more adaptable and sturdy than this one from Sculpture Depot. It's the first, and last, stand you'll need to buy.

SCULPEY MEDIUM
$10: SCULPEY.COM
Hallelujah! No more mixing soft and hard to get that perfect blend. I think I cried a little when I got my first brick of this stuff. It still bakes perfectly to a rock-hard state and sands beautifully.

MONSTER CLAY
$30 FOR 5 LBS: MONSTERMAKERS.COM
This sulfur-free, oil-based smooth clay has replaced almost all other clays for me. I can crank out detail and organic shapes, yet it holds up to sanding and smoothing better than anything I have used. Monster Makers will soon offer hard and soft versions — I can't wait.

YASUTOMO NIJI ROLL
$12: YASUTOMO.COM
My favorite, no frills tool wrap. What makes this so special? Most sculpting utensils have different tools on each end. This wrap shows off both ends, instead of the regular pocket design, allowing a fast glance for that perfect tool you are looking for.

KEN'S TOOLS ITTY BITTY 1
$13.75: KENSTOOLS.COM
Ken Banks' Itty Bitty line shines among his rake and loop tools. They hold up to abuse on hard wax detailing or large clay raking. These absolutely never leave my tool wrap.

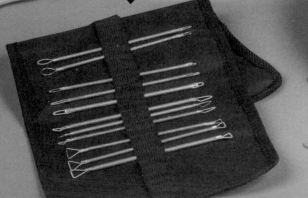

Hep Svadja

Written by Jason Babler

MAKE: BELIEVE

makezine.com/go/ gift-guide-2015

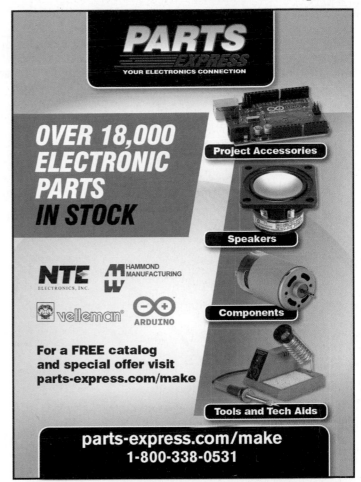
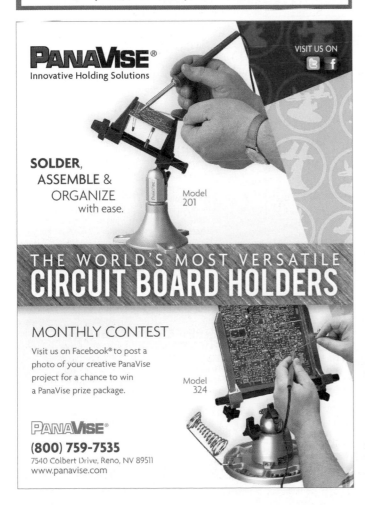
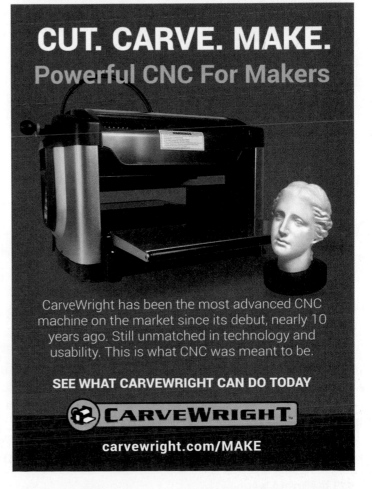

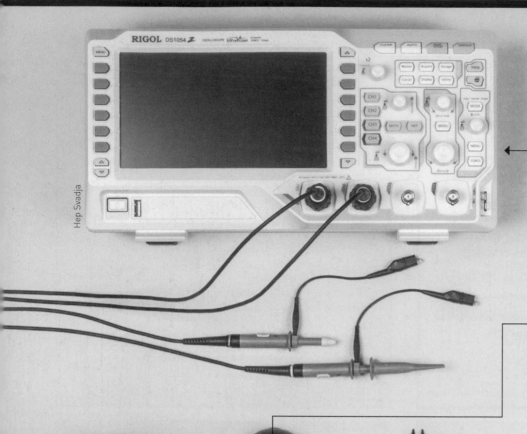

Hep Svadja

RIGOL DS1054Z DIGITAL OSCILLOSCOPE
$399: RIGOLNA.COM

If you're doing any reverse engineering or troubleshooting of complex circuits, you'll definitely want an oscilloscope. The Rigol DS1054Z is a 50MHz digital scope that captures up to a gigasample per second, and is the most affordable 4-channel scope available. If you're investing in an oscilloscope, this one will do just about everything the average hobbyist needs.

HAKKO FX-888D SOLDERING STATION
$107: HAKKOUSA.COM

The FX-888D is a good general purpose soldering station that doesn't break the bank. It can handle almost anything you'll throw at it, and Hakko has many different affordable tips available for specialty work. The station also comes with multiple preset temperatures for quick adjustment. Bundle option includes wire cutter.

MG CHEMICALS 400NS SERIES #3 SUPER WICK
$5: MGCHEMICALS.COM

Desoldering wick is the unsung hero of board repair. This wick contains flux to help solder flow into it, and any residue left behind is nonconductive and noncorrosive. It's a must-have for anyone doing circuit board repair or rework.

GETTING STARTED WITH ARDUINO KIT V3.0
$65: MAKERSHED.COM

For anyone getting started in the world of microcontrollers, look no further than the *Make: Getting Started with Arduino* kit. It includes an Arduino UNO, breadboard, jumper wires, LEDs, resistors, photoresistors, buttons, and associated cables to get you up and running as quickly as possible. It's a great kit for beginners, and a great way to get into electronics.

ENGINEER SS-02 SOLDER SUCKER
$23: ENGINEER.JP

This compact solder sucker easily fits in a portable kit, and it outperforms any other solder sucker I've used. Rather than rigid teflon like most other suckers, its tip is made from flexible, heat-resistant silicone tubing. This lets you push directly on top of the soldering iron tip for maximum desoldering suction.

Written by Jordan Bunker

ELECTRONICS

makezine.com/go/gift-guide-2015